Logic and Philosophy of Mathematics in the Early Husserl

SYNTHESE LIBRARY

STUDIES IN EPISTEMOLOGY, LOGIC, METHODOLOGY, AND PHILOSOPHY OF SCIENCE

VOLUME 345

For further volumes:
http://www.springer.com/series/6607

Logic and Philosophy of Mathematics in the Early Husserl

by

Stefania Centrone
University of Hamburg, Germany

Springer

Dr. Stefania Centrone
Universität Hamburg
Philosophisches Seminar
Von-Melle-Park 6
20146 Hamburg
Germany
stefania.centrone@uni-hamburg.de

ISBN: 978-90-481-3245-4 e-ISBN: 978-90-481-3246-1
DOI 10.1007/978-90-481-3246-1
Springer Dordrecht Heidelberg London New York

Library of Congress Control Number: 2009941297

Printed on acid-free paper

9 8 7 6 5 4 3 2 1

Springer is part of Springer Science+Business Media (www.springer.com)

Foreword

Early Husserl, Mathematics and Logic

Edmund Husserl's historically inalienable role as "the father of phenomenology" and the attitudes this description arouses in his friends and foes alike have led to a persistent and systematic disregard of his early work. Where notice is taken of it at all, it is generally considered as a product of apprenticeship, while he was learning his trade, before the breakthrough work of the *Logical Investigations* and the methodological turn to phenomenology with its attendant reductions and transcendental idealism. Husserl began his career as a mathematician, so the line tends to be that it was natural for him to start there but at least as natural for him to move on to the bigger (one might say, "more grown-up") issues of the foundations of logic and methodology in general. Certainly those admirers and detractors of Husserl who see his main role as a progenitor of so-called Continental philosophy are likely to be both less attuned to the interests of a philosopher who had more in common with Frege and Hilbert than with Heidegger and Derrida, and less inclined to accord that background a role in appraising Husserl's contribution to thought.

Stefania Centrone's thorough and painstaking exposition of Husserl's early work is a timely reminder that he was a philosopher of insight and stature well before he burst onto the general philosophical scene. One would hope this realisation could become universal, but given entrenched interests and attitudes it is unlikely to be heeded as widely as it should. So why should we take note of the early Husserl? What relevance does it have for his own development, and what significance for philosophy at the turn of the twentieth century and beyond?

The history of the philosophy of mathematics during the golden years of 1879–1939 hardly ever mentions Husserl. One reads about Dedekind, Cantor, Frege, Peano, Russell, Poincaré, Hilbert, Brouwer, Weyl, Gödel, Church and Turing. Husserl is effectively written out of the picture because his *Philosophie der Arithmetik* was criticised as psychologistic by Frege, and majority opinion is on Frege's side. Husserl, goes the story, saw the light, realised Frege was right, rounded on the psychologism of his earlier self and his teacher Brentano, did his penance, and then swiftly moved on to other things, turning to the more exciting and ultimately more popular forms of essentialism and idealism that stimulated two generations of students in Göttingen and Freiburg. The fourteen "lost years" in Halle are consigned to prehistory. It is my contention, based in part on the compelling evidence presented in Stefania Centrone's book, that Husserl would deserve an honourable mention in the history of the philosophy of mathematics and logic alongside the others, and that this would indeed have been more apparent had he not gone on to become the philosophical colossus with whom we are familiar.

Firstly, and for the record, Husserl was the first person outside Jena to take Frege seriously. This is despite the fact that Husserl's Halle colleague and friend Georg Cantor knew a little about Frege from an early and very sketchy review of Frege's *Begriffsschrift*. In later life, Husserl rather cruelly described Frege to Heinrich Scholz as an oddball (*Sonderling*), which was at the time an accurate reflection of the general perception of Frege's role and status. Nevertheless, unlike anyone before Russell, Husserl early on paid Frege the compliment of reading *Die Grundlagen der Arithmetik*, thinking about its theory, and criticising it in his own first book *Philosophie der Arithmetik*. And in two respects at least, Husserl's criticisms are right on the money. The first is that Frege's choice of the extension of the concept "equinumerous with the concept *F*" to be the number of the *F*s, is clearly artificial, and not what we understand by number. Secondly, Husserl's analysis and ontology of number are preferable to Frege's. We take number to be neither a property of concepts nor an abstract object but a non-distributive formal property of collections. 'Four' is not a property of the concept "evangelist", nor is it an object which is the extension of a second-order concept, but a property of the group or collection of the evangelists, the four of them. Husserl does go too far in criticising Frege in that he sees no role whatsoever for the idea of abstraction under an equivalence relation of equinumerosity. That most useful insight of Frege can and should be coupled to Husserl's multitude theory of numbers. So he is not an infallible guide to Frege. But Frege is a worse guide to Husserl. Having spotted in Husserl a number of statements that are construable psychologistically, Frege took the opportunity in a review of *Philosophie der Arithmetik* to pay Husserl back with compound interest for the temerity of having criticised him. Some of Frege's barbed and ill-tempered criticism is accurate, but a lot of it is not. So the legend was born that Husserl was converted by the strength and cogency of Frege's criticisms into being a crusading anti-psychologist himself. Certainly Husserl recognised that Frege had made some valid points, and was grateful for the criticism, but it is one-sided to suppose he was not already becoming dissatisfied with aspects of his early philosophy of arithmetic, aspects which stopped him from completing the planned second volume. On the other hand Husserl's over-zealous defenders have insisted that Frege had nothing whatsoever to teach him and that his changes of mind were wholly intrinsic and independent of the criticism. The truth lies somewhere in between. But the fact remains that Husserl never changed his view as to the nature of natural numbers as properties of collections or multitudes, so on the substantial issue of the correct ontology of arithmetic, he was unmoved, and I think rightly so.

The formative influence on Husserl's development as a philosopher was of course Franz Brentano, as Husserl was happy to acknowledge, and as he demonstrates in the dedication of *Philosophie der Arithmetik*, a dedication incidentally, which Brentano only belatedly and grudgingly acknowledged. But the Brentano who influenced Husserl was the Brentano of the lecture theatre, not the Brentano of the published works. So the impression could easily arise that all the mathematical ideas in *Philosophie der Arithmetik* were Husserl's. In fact the idea of collective combination is already present in rudimentary form in the 1884/5 Vienna lectures on logic that Husserl attended, though no acknowledgement finds its way into Husserl's text, perhaps because the source was unpublished. A balanced and objective judgement of the extent to which Husserl's ideas at this stage are indebted to those of Brentano must await a proper publication of Brentano's chaotic *Nachlass*. Husserl's work on the philosophy of mathematics from the 1890s was also largely unpublished at the time, but his literary remains have received a much more complete and favourable treatment than those of Brentano, so it is possible to examine in print the many manuscripts, notes and lecture notes from this period. They show him to have been interested in a wide variety of topics in the philosophy of mathematics and logic, and to have anticipated a number of topics that later became common currency in ensuing years. Stefania Centrone shows that some of these ideas, in particular those connected with one concept of completness, and the use of ideal or "imaginary" elements in formal mathematics, became key concepts in the Hilbertian formalism programme, without receiving from Hilbert the due they deserved. There may have been personal reasons for this: Hilbert was in favour of Husserl's appointment at Göttingen, hoping to find a philosopher to whom he could talk philosophy of mathematics, but Husserl was already broadening his interests in other directions, and they fell out over university matters, in particular the promotion of Leonard Nelson.

The most important influence on Husserl the logician was undoubtedly that of Bernard Bolzano, whose mighty *Wissenschaftslehre* (1837) Husserl chanced upon in a second-hand book shop. The *Wissenschaftslehre* anticipated or indeed forged many ideas still now widely regarded as achievements of later logicians such as Frege, Tarski, and Quine. But for Husserl's serendipitous find, Bolzano's genius might well have lain undiscovered for much longer, as Husserl is lavish with his praise in the *Logical Investigations*. Bolzano's semantic platonism was congenial to Husserl, and was turned to use as the objectivistic alternative to psychologism in logic, while being modified by Husserl to conform with Brentano's intentionality theory of the mental. Husserl was also pleased to adopt and adapt technical concepts from Bolzano's logic, explained in semantic terms, which he had not taken from Frege, whose philosophical outlook precluded any idea of semantics as a proper logical discipline. Husserl, having absorbed his formal logic initially from the logical algebraist Ernst Schröder, was no stranger to semantic matters, indeed he proposed an intensionalistic reading of Schröder's inclusion and defended this against the extensionalist logician Voigt. Husserl was never one interested in investigating logical calculi for their own sake: he was interested neither in proving theorems nor in seeing how axiomatic systems could be refined, manipulated, and simplified. Most of his remarks are metalogical rather than logical, and this goes for his remarks on mathematics as well. It is possible to see in Husserl's instinctive practice the seeds of the later concept of metamathematics of Hilbert. At the same time Husserl's lack of interest in formal manipulation stood him in poor stead when it came to gaining the respect and citation of mathematicians. Nevertheless, while he did not venture into such areas in his published writings, Husserl was quite prepared in lectures to work through logical proofs in the modern manner, as Stefania Centrone shows with regard to the logic lectures of 1896. In her chapter on Husserl the logician, Stefania Centrone steers us through the complications of Husserl's attitude to logic, his borrowings, modifications and influences.

It has long been known that during Hilbert's first phase of encounter with the foundations of mathematics, around the turn of the twentieth century, he was interested in the question whether all mathematical questions have a definite yes or no answer, and whether mathematicians can in principle show what the answer is. It has also been noted that Husserl also thought about such matters, though much of the evidence about this came from his later work *Formal and Transcendental Logic* of 1929, which represents his last published writings on the foundations of mathematics and logic. In her final chapter Stefania Centrone shows that Husserl's thinking originated much earlier, in a lecture of 1901 delivered shortly after his arrival in Göttingen, and likely to have resonated strongly with Hilbert. This lecture shows Husserl to be fully aware of and indeed himself advancing cutting-edge developments in the philosophy of mathematics: issues of formalization, algebraicization, decidability, completeness, models, and the consistent extension of mathematical concepts into new mathematical systems. If Husserl's ideas seem inchoate and unfocussed by today's standards, it is instructive to compare them with Hilbert's own writings, of this time and indeed later. To those accustomed to the limpid clarities of Frege and Russell, Hilbert's writings, despite their evident suggestiveness, and their origin in a mathematician of world ranking, are at times alarmingly unclear, and were not decisively sharpened until much later with the help of Paul Bernays and others. By those standards, Husserl's writing is equally suggestive and no less clear. A comparison with the 1929 work shows little subsequent advance. This is unsurprising, since Husserl, unlike Hilbert and others who returned again and again to the problems of logic and the foundation of mathematics, was preoccupied with many other philosophical matters after 1901, leaving it to others to gain the laurels for work on the foundations of mathematics.

Stefania Centrone's book presents us with an aspect of Husserl that, under counterfactual circumstances, might have been the familiar one: Husserl the innovative and thoughtful philosopher of mathematics and logic. It is instructive to engage in a little *epoché*: bracket the familiar Husserl of intentionality, phenomenology and transcendental idealism, and consider the colleague and contemporary of Cantor and Hilbert, writing about sets, numbers, consistency, formalization and the domains of theories. There is still much to learn about this phase and aspect of Husserl's thought, but thanks to Dr Centrone, it is now a good deal easier to engage with that enterprise.

Peter Simons
Trinity College Dublin

Preface

This book has been long in making. It partially originates from my doctoral dissertation that I wrote under the supervision of Professor Ettore Casari at the Scuola Normale Superiore in Pisa and defended in 2004. Further research on the main topics of the present book at the SNS in Pisa was supported by a 2 year post-doc grant for research on *Themes and problems of logical objectivism*, again under the supervision of Casari, followed by a 18 months research grant on *Logic and Philosophy in Husserl*, supervised by Professor Massimo Mugnai. I could finally complete this book in the hospitable environment of the Philosophy Department of Hamburg University, since Professor Wolfgang Künne, my present supervisor, kindly allowed me to use for this purpose the first 2 months of a 2 years Alexander-von-Humboldt fellowship for research on *Logical Objectivism, Inference and Foundational Proof in Bernard Bolzano's 'Wissenschaftslehre'*.

I am particularly indebted to my first teacher, Ettore Casari, who aroused my interest in logic, in mathematics and in Husserl's early writings and who made me realize that, as he used to put it, *"il mondo è vasto"*, i.e. that restriction to one single field of research can be more of a hindrance than a help for original work. Very special thanks must go to Professor Kevin Mulligan of the University of Geneva. We first met on the occasion of the defense of my PhD thesis. From then on he encouraged me more than anyone else in broadening and deepening the research I had begun in my thesis. We continued talking on many of the topics of this book over the years, and hopefully a trace of these discussions will be visible in many pages of my work. Many thanks go to Massimo Mugnai, who strongly supported my work and instilled in me an admiration for Leibniz that is bound to last, and to Burt Hopkins who very much encouraged the realization of my project. I would also like to thank Claudio Cesa and Francesco del Punta for many hours of enlightening discussions about the history of ideas. I am especially indebted to Wolfgang Künne. He commented incisively on every chapter, insisted on many clarifications and saw to it that I became more aware than ever of the importance of many issues in Bolzano's still sadly neglected Logic and in what Wittgenstein praised as "*die großartigen Werke Freges*".

Much feedback I had over the years from many scholars, in particular: Sergio Bernini, Arianna Betti, Riccardo Bruni, Andrea Cantini, Laura Crosilla, Carlo Ierna and Francesca Poggiolesi. The criticisms and suggestions made by an anonymous referee for Synthese Library who read the penultimate draft of this book were very helpful. The encouragement at a decisive moment and the friendly advice I received from Willemijn Arts, the Senior Publishing Editor, and from Ingrid van Laarhoven, the Senior Publishing Assistant of Springer Science and Business Media, were truly invaluable. Very special thanks must also go to Maja de Keijzer, Publishing Editor.

I am very grateful to my mother Nicoletta who made me love hard work and to my father Mario who aroused in me the love for philosophy. – With gratitude and affection I dedicate this book to my husband and best friend Piero.

Introduction

This book takes into account the first ten years of Edmund Husserl's work, from the publication of the *Philosophy of Arithmetic* (1891) to that of the *Logical Investigations* (1900/01), with the aim of precisely locating his early work in the field of logic and the philosophy of mathematics. This goal does seem to be worth pursuing especially in the light of the developments in formal logic during the past century. Surveying the vast growth of studies on this topic since the second world war, a tendency can be seen to emerge among the interpreters of Husserl's thought to remain within the methodological and even terminological bounds of Husserl's later phenomenology while, conversely, professional logicians fail to consider Husserl's contributions to the field of formal logic as significant for their discipline.

Our decision to focus upon Husserl's early reflections on logic and the philosophy of mathematics and to consider only selectively their elaboration in his mature work is motivated by the fact that these ideas were definitely original and surprisingly innovative at the moment of their first conception, i.e., in the years 1896–1901 when Husserl worked on the *Prolegomena*, while they no longer appear as fresh (though they are sometimes better articulated and corroborated) when they are taken up again in *Formal and Transcendental Logic* (1929). These ideas include, to mention some significant examples, the articulation of formal logic in logical levels according to a structure that is very close to what, today, is effectively used in standard logical textbooks, the unification of formal logic and mathematics in a most general mathematico-formal science that purports to be the concrete realization of the Leibnizian ideal of a *mathesis universalis*, and the explicit conception of abstract mathematics as a theory of structures.

The goal of our work is to restore the level of the real discussion between Husserl and his important early interlocutors, some of whom made definitive contributions to the development of formal logic as an autonomous discipline in the last two centuries. To this end we will consider Husserl's relationship to the algebraists of logic, in particular George Boole, as well as to Bernard Bolzano's, Gottlob Frege's and David Hilbert's contributions to logic.

With respect to the two main possibilities for textual research, philological and erudite commentary on the one hand, and comprehensive interpretative stances, on the other hand, this book opts for the second. Its contributions are almost exclusively analytical and, on the basis of a close reading of selected texts written in the indicated decade, it aims to bring to light the unity and depth of an original and comprehensive design of both a *theoretical systematization* of logic and its *philosophical foundation.*

The distinctive trait of Husserl's work during the period in question is the simultaneous presence in his logical and mathematical reflections of two different directions of research, (1) the project of a substantial mathematization of logic and (2) a conception of logic as the study of objective relations occurring among certain abstract logical entities. As regards (1), we find Husserl's interest in specifically *logico-formal* issues: he succeeds in grasping with great clarity and insight the implications of the *formal-abstract* trend in mathematics and, in particular, of its tendency toward *algebrization*, which he is able to transfer to and elaborate at the logico-theoretical level. As regards (2), we find Husserl's project to develop a philosophy of logic and mathematics focused on the systematic investigation of the properties and relations that occur among certain abstract semantical entities: a source of inspiration for this project is the theory of *Notions* (*Vorstellungen an sich*) and *Propositions* (*Sätze an sich*) in Bolzano's *Wissenschaftslehre,* and one of its more remote ancestors is the Stoic doctrine of *Sayables* (*lektá*). In this book we shall mainly focus on the research direction (1).

In **Chapter 1** we take Husserl's first major work, the *Philosophy of Arithmetic* (1891), as the starting point of our study. Dagfinn Føllesdal's conjectured in 1958 that Frege was an important factor in Husserl's conversion from the psychologism of this book to the anti-psychologism of the *Prolegomena.* This claim has been contested by Mohanty and others, but Føllesdal's defense is very convincing.[1] However, we will approach Husserl's first book from a perspective that is orthogonal to the psychologism issue. Rudolf Bernet has written that the *Philosophy of Arithmetic* "represents, not a mere youthful transgression stemming from Husserl's psychologistic period, but a highly valuable work of intrinsic and enduring importance". According to Bernet its value lies in the fact that "the *Philosophy of Arithmetic* . . . anticipates certain decisive results not only of the *Logical Investigations* but also of Husserl's later work."[2] On our view, however, the value of this text exceeds that of anticipating some claims that came to be consolidated in Husserl's phenomenology. The specific solutions that Husserl advances in his first book possess an intrinsic interest for logic and mathematics, and they are independent of the psychologistic context in which they originate.

In his *Philosophy of Arithmetic* Husserl enters into a very lively and stimulating debate about the foundational issues regarding the concepts of *number* and *set.* Moreover, this work contains many interesting insights regarding the formal and

---

[1]D. Føllesdal 1958 (1994) and 1982, a reply to one of his critics.

[2]Bernet & Kern & Marbach 1989, 14.

computational aspect of theories. For instance, we find the elaboration of the concept of "number system" and investigations aiming at circumscribing the totality of all 'conceivable arithmetical operations', which bring to light how Husserl had arrived at a first delimitation of the class of number functions that today are called "partial recursive functions." In this context we discuss a part of Husserl's *Nachlass* text "On the Concept of Operation" (also from 1891) which develops a specifically logico-formal problem raised in the *Philosophy of Arithmetic* concerning the question of the formal irreducibility of the operation of multiplication to that of addition. We shall also discuss Husserl's relationship to Boole as regards the conception of the more properly formal and calculatorial aspect of theories and his relationship to Frege as regards the definition of the series of natural numbers. With respect to the latter problem we also take into consideration a *Nachlass* text "*On the Theory of Sets*" which is centered on the distinction between *finite* and *infinite* cardinals: according to Husserl himself, it was a grave "defect" of the *Philosophy of Arithmetic* not to have provided a theoretical account of this distinction. The heart of the issue is this: Husserl had defined natural numbers as the collection of all those objects that can be obtained starting from *zero* using a *finite number* of steps to *successors*, but he had not registered the fact that the crucial point of such a definition is precisely to reformulate successfully the reference to a "finite number of successor-steps" *without using the concept of a finite number* (since to use the latter concept is to fall into a vicious circle). Husserl's account is therefore at variance with what Frege had already done informally in his *Grundlagen* (1884) and then formally in the *Grundgesetze* (1893).

Chapter 2 on "the idea of a pure logic" examines selected themes belonging to the philosophy of mathematics and logic frequently raised and discussed by Husserl in the years between 1896 and 1900. The discussion pivots on various issues connected to the surprisingly innovative idea of a *stratification of formal logic* in logical levels. Roughly, (1) he outlines what was to become the modern conception of a formal language (*logical morphology*), (2) he provides a sketch of a propositional logic and a quantified logic (later in *Formal and Transcendental Logic* called "logic of consequences"), and (3) he largely anticipates the modern concept of a formal system (*theory of theories*). In this context, his attempt to unfold a concept of semi-formal enthymematic derivability and to characterize a notion of "dependency among truths," i.e. of a one-way entailment (between true propositions) of a reason-giving kind, plays a prominent role. Hence we have to consider the relation between Bolzano's notions of derivability (*Ableitbarkeit*) and consecutivity (*Abfolge*) and Husserl's notions of 'following from certain premises through correct inferences' and of 'grounding' or 'foundation' (*Begründung*).

At this point, a few words are in order about the importance of Bolzano's monumental *Wissenschaftslehre* for Husserl's early work, say from 1893–94 onwards. Husserl himself finds it important to stress in an appendix to Chapter 10 of the *Prolegomena*[3] that his investigations are not "in any sense mere commentaries upon, or critically improved expositions of, Bolzano's thought patterns", but that they "have been crucially stimulated by Bolzano ...". In Husserl's eyes, Bolzano's great merit lies in his characterizing pure logic as a discipline that is concerned "with the most general conditions of *truth* itself"[4] and deals with the relations among the *contents* of our thoughts. So the emphasis is on Bolzanos' logical objectivism. He praises Bolzano's *opus magnum* as "a work which .... far surpasses everything that world-literature has to offer in the way of a systematic contribution to logic"[5]:

> Bolzano did not, of course, expressly discuss or support any independent demarcation of pure logic in our sense, but he provided one *de facto* in the first two volumes of his work, in his discussions of what underlay a *Wissenschaftslehre* or theory of science in the sense of his conception; he did so with such purity and scientific strictness, and with such a rich store of original, scientifically confirmed and ever fruitful thoughts, that we must count him as one of the greatest logicians of all time... Logic as a science must ... be built upon Bolzano's work, and must learn from him its need for mathematical acuteness in distinctions, for mathematical exactness in theories. It will then reach a new standpoint for judging the mathematizing theories of logic, which mathematicians, quite unperturbed by philosophic scorn, are so successfully constructing.

However, Husserl directs at Bolzano two sorts of criticism[6] which are worth to be mentioned already in this Introduction. *Firstly*, though having circumscribed the domain of pure logic as "a closed field of independent and *a priori* abstract truths", Bolzano sees his investigations in the service of a science which sets up "the rules according to which we must proceed in the business of dividing the entire realm of truth into single sciences and in the exposition thereof in special textbooks".[7]

[3]*PR* 224–227 (*Hinweise auf F.A. Lange und B. Bolzano*), *Pre* 222–224. The quotations that follow are all taken from this passage. Henceforth: *PR* = Husserl, *Logische Untersuchungen* I, *Prolegomena zur reinen Logik*, Tübingen 1993; *PRe* = English translation thereof, in: *Logical Investigations*, London 1970, Vol. I, 51–247. Responsibility for translations from German is mine, even when I refer to, benefit from, or simply echo published translations.

[4]Bolzano, *Wissenschaftslehre* (Sulzbach 1837), I, §16, 65. Henceforth: *WL*.

[5]In 1911 the key is a bit lower: these volumes, he now says, occupy "the highest rank in the logical world-literature of the 19th century" (quoted in Künne 2009, note 1, and commented upon in his 2008, 358).

[6]"Much as Bolzano's achievement is 'cast in one piece', it cannot be regarded (as such a deeply honest thinker would be the first to admit) as in any way final."

[7]Bolzano, *WL* I, §1, 7. In a note to Chapter 1 of the *Prolegomena* Husserl writes "The fourth volume of the *Wissenschaftslehre* is indeed especially devoted to the task which the definition expresses [The theory of science (or logic) is "the science which shows us how to present the sciences in convenient textbooks"]. But it strikes one as strange that the incomparably more important disciplines which the first three volumes treat of, should be represented merely as aids to a technology of scientific textbooks. Naturally, too, the values of this by no means as yet sufficiently valued work, which is, in fact, almost unused, rests on the researches of these earlier volumes" (*PR* 29, *PRe* 73).

According to Husserl, this relation should be inverted: pure logic should ground logic as a practical discipline. *Secondly*, Bolzano "did not quite exhaust the rich inspiration of Leibniz's logical intuitions, especially not in regard to mathematical syllogistics and to *mathesis universalis*". This is a criticism that Husserl resumes in more detail in his *Formale und transzendentale Logik*.[8] One may very well wonder whether it really hits its target, but it is of great significance because it highlights Husserl's attitude towards Bolzano's project of a unification of logic and mathematics in a most comprehensive science. In § 8 of Part I of his *Beyträge zu einer begründeteren Darstellung der Mathematik* (1810)[9] Bolzano defines mathematics as "a science which treats of the universal laws (forms) things must comply with in their existence (*eine Wissenschaft, die von den allgemeinen Gesetzen (Formen) handelt, nach welchen sich die Dinge in ihrem Daseyn richten müssen*)", where 'thing' is meant to cover "everything that can be object of our representational capacity". In I, § 9 he says of mathematics that it is "concerned with the question: what must things be like if they are to be possible at all (*wie müssen die Dinge beschaffen seyn, die möglich seyn sollen?*)" And in I, § 11 he says that the laws of what he calls "*die allgemeine Mathesis*" are "applicable to all things without any exception (*auf alle Dinge ganz ohne Ausnahme anwendbar*)". This discipline comprises, inter alia, *Logistik oder Arithmetik* and *Combinationslehre* (cp. I, § 3), whereas disciplines like geometry and chronometry are "subordinated to the whole universal mathesis as species to a genus (*der allgemeinen Mathesis insgesammt, wie Arten der Gattung, subordinirt*)".[10] Now Husserl acknowledges that Bolzano characterizes here "a universal *apriori* ontology", but he objects that Bolzano does not develop all features of *formalization*, of the transition from the material to the formal, and that he fails to keep the formal and the material aspects of ontology clearly distinct.

> When he conceives of the thing as such (*Ding überhaupt*) as the highest genus . . . it becomes clear that he did not see the difference between the empty form of the something as such as highest genus . . . and the universal realm of possible existents, of the real in the widest sense (*die universale Region des möglicherweise Daseienden, des im weitesten Sinne Realen*)), which differentiates itself in particular regions. He also did not see the difference between the subsumption of formal particularities (*Besonderungen*) under formal generalities and the subsumption of regional particularities . . . under formal generalities. . . In other words, Bolzano did not attain the proper concept of the formal,. . ., though he touched it somehow.[11]

In connection with Husserl's reflections upon the idea of a pure logic, we shall also discuss in Chapter 2 his development of a propositional calculus of the axiomatic-deductive kind, which is found in the final section of a lecture course on logic held at the university of Halle in summer 1896. This lecture course,

---

[8]*Formal and Transcendental Logic* [ed. 1929] (henceforth cited as *FTL*) 74–75.

[9]See below, Ch. 2, § 2, n. 32.

[10]Cp. Casari 2004, 161.

[11]*FTL* 74–75.

generally known as *Logikvorlesung 1896*,[12] contains part of the core reflections that gave rise to the *Prolegomena* and the *Logical Investigations*.

It does *not yet* give us the "*Grundgerüst* (the basic scaffolding)" of the *Prolegomena*, in spite of Husserl's claim to the contrary in the *Preface* to the second edition of the *Logical Investigations*.[13] The editor of the critical Husserliana edition, Elisabeth Schuhmann, notes in her Introduction that there is an error in Husserl's own dating. He did not hold two complementary courses in 1896. Moreover, only a few pages of the *Prolegomena* (more exactly, §§ 4–8) coincide with material in the *Logikvorlesung*. The reason for this error is probably the fact that Husserl repeatedly re-used material from the *Logikvorlesung* of 1896 to prepare additional logic courses, for instance, the course "*Logik und Erkenntnistheorie*" (winter term 1901/02), the courses "*Logik*" and "*Allgemeine Erkenntnistheorie*" (winter 1902/03), and the series of lectures "*Alte und neue Logik*" (winter 1908/09). In particular, manuscripts from 1901/02, written after and based on the *Prolegomena*, were collected together – without indications of the date – with the *Logikvorlesung* of 1896. When preparing the new edition of the *Logical Investigations* in 1913, Husserl must have found them in the same 'convolute' as the 1901/02 lectures on "*Logik und Erkenntnistheorie*", which were also undated. Hence, in the draft for the preface to the new edition of the *Logical Investigations*, he wrote that the *Prolegomena* were, essentially, only an elaboration of the *Logikvorlesung* of the summer and winter 1896.

The issue of *imaginary numbers*, and, more precisely, of the "*logical meaning of the calculatory passage through the imaginary*," which is, without doubt, the guiding thread in Husserl's reflections on the role of the *formal* attitude in mathematics, is the specific topic of his famous "double" lecture (known as the *Doppelvortrag*) presented to the *Mathematische Gesellschaft* in Göttingen in winter 1901.

Chapter 3, the final chapter of this book, is focused on this and other thematically related texts of the *Nachlass*. In particular, we will emphasize Husserl's reflections on the notion of a *formal theory* in its double aspect of a *system of axioms* and the *manifold* underlying it. We will focus, furthermore, on the more specific reflections regarding, on the one hand, the structure of (what Husserl calls) *Universal Arithmetic* – i.e., a system of calculation rules valid in all number-systems (cardinal numbers, whole numbers, etc.), and, on the other hand, the structure of the *specific Arithmetics* or *systems of operations* – i.e. systems of calculation rules that contain those of universal arithmetic as common part plus some specific groups of rules able to characterize the behavior of arithmetical operations relating to a specific number system. Finally, we will consider Husserl's

[12]Husserl, *Logik. Vorlesung 1896*, ed. Elisabeth Schumann, Husserliana Materialienbände, Band 1, Kluwer, Dordrecht 2001. Henceforth: *LV'96*.

[13]"The *Prolegomena to Pure Logic* are, in their essential content, a simple elaboration of two complementary lecture courses held in Halle in the summer and winter of 1896."

reflections on the fundamental and closely connected notions of the *definiteness*, and the *formal extension*, of a theory.

In his many references to the *Doppelvortrag* (inter alia, in the second edition of the *Prolegomena*, in the *Ideas* and finally in *Formal and Transcendental Logic*) Husserl observes that some important ideas which he presented on that occasion were subsequently taken over, without acknowledgement, in the logical investigations of Hilbert's school.

> The concepts introduced here [Husserl means specifically the concept of a *definite* system of axioms] served me already at the beginning of the 1890s (in the "*Untersuchungen zur Theorie der formal-mathematischen Disziplinen* [Investigations pertaining to the theory of formal-mathematical disciplines]", which I intended as a continuation of my *Philosophie der Arithmetik*), to find a *fundamental* solution to the problem of the imaginary. ... Since then I have often had occasion to develop the relevant concepts and theories in lectures and seminars, partly in complete detail; and in the winter semester of 1901/02 I dealt with them in a double lecture to the Göttingen Mathematical Society. Some parts of this train of thoughts have found their way into the literature, without mention of their original sources. – The close relationship of the concept of definiteness to the "axiom of completeness" introduced by Hilbert for the foundation of arithmetic will be immediately obvious to every mathematician.[14]

In his *Doppelvortrag* Husserl examines two notions of definiteness: "absolute definiteness," which, as he indicates, is analogous to Hilbert's axiom of completeness, and "relative definiteness," which he applies to systems of axioms and to the structures that underlie theories conceived of as deductive systems. Basically, a system of axioms that is "definite in the absolute sense" or "in the Hilbertian sense" (as Husserl puts it) is categorical, i.e. it individuates, up to isomorphism, only one model, whereas a system of axioms that is "definite in the relative sense" is not necessarily categorical, but it is such that every proposition written in the language of the theory can be decided on the basis of the axioms. Given the different implications of these two distinct notions, the aim of giving a rigorous (mathematical) definition seems to be worth pursuing. It should help to weed out some common misconceptions as regards the interpretation of these issues and to challenge some recent and well-documented contributions to this topic.

It is worth emphasizing that Husserl himself has pointed out that "the progress from vaguely formed, to mathematically exact, concepts and theories is, here as everywhere, the precondition for full insight into *a priori* connections and an inescapable demand of science".[15]

---

[14]Husserl, *Ideen I*, § 72, n.1; *Ideas* 164, n. 17.

[15]Loc. cit.

Contents

Chapter 1
Philosophy of Arithmetic

1.1 Introduction

The *Philosophy of Arithmetic*,[1] Husserl's youthful work dedicated to a philosophical, or better, epistemological foundation of mathematics, shows the shift in his interests from more properly mathematical issues to those regarding the *philosophy of mathematics*. Husserl strives to understand and clarify *what* numbers and numerical relations *are*, a problem that he recasts in terms of the *subjective origin*[2] of the fundamental concepts of set theory and finite cardinal arithmetic. We will try to show that on the whole this work of Husserl's does not deserve the criticism and ensuing neglect that it suffered from, ever since Frege published his well-known *Review*.[3] Besides its hotly contested psychologism, we find ideas and conceptualizations that not only were original then, but are still interesting today, such as those concerning the autonomy of the formal-algorithmic aspect of abstract algebra and mathematics. Moreover, it is here that the Husserlian idea of a *universal arithmetic* receives its first formulation, the full elaboration of which will take at least ten more years, until his research on these topics reaches its stable form in 1901.[4]

[1]Husserl, *Philosophie der Arithmetik*. Mit ergänzenden Texten (1890–1901), Huss XII, 1–283. Henceforth cited as *PdA*. English translation cited as *PoA*.

[2]Cp. Tieszen 1996: "Husserl thinks that arithmetical knowledge is originally built up in founding acts from basic, everyday intuitions in a way that reflects our a priori cognitive involvement" (304).

[3]Frege 1894. Cp., for example: Osborn 1949; Picker 1962, 289; Beth 1966, 353. Among the interpretations that give a positive re-evaluation of some aspects of the *PdA*: Farber 1943; Føllesdal 1958; Haddock 1973 (especially Ch. VI: however, his focus is mainly on Husserl's logical theories in his later works, in particular in the *Logical Investigations* and *Formal and Transcendental Logic*); Miller 1982; Willard 1974, 97 f. & 1984; Tieszen 1990; Ortiz Hill 1994a & b.

[4]See *Das Imaginäre in der Mathematik* (December/January 1901/02), *PdA*, App. 430–451, *PoA* 409–432, and the new critical edition Schumann & Schumann 2001. Willard 1984 rightly stresses that Husserl shared "the general persuasion of mathematicians of the time that a rigorous development of higher analysis – *arithmetica universalis* in Newton's sense – would have to emanate from elementary arithmetic alone." However, few lines later he writes that "these further

S. Centrone, *Logic and Philosophy of Mathematics in the Early Husserl*,
Synthese Library 345,
DOI 10.1007/978-90-481-3246-1_1,

Of the two volumes of which the work, according to Husserl's initial plan, was to consist, only the first was completed and published (Halle 1891). In spite of the preliminary nature of the studies intended for the second volume that we possess,[5] we know that it was to contain two parts: one dedicated to "the justification of utilizing in calculations the quasi-numbers (*Quasizahlen*) originating out of the inverse operations: the negative, imaginary, fractional and irrational numbers,"[6] the other to the determination of the general characteristics of a universal arithmetic.

In the first and only volume the contents of the *Habilitationsschrift "On the Concept of Number: Psychological Analyses*" (Halle 1887)[7] appeared as Chapters I–IV without significant changes. It had as its main topic the constitution of the concept of cardinal number (*Anzahl*), and it also consists of two parts. The first part studies the fundamental concepts of mathematics – *multiplicity* (*Vielheit*), *cardinal*

---

matters – intended is the foundation of the whole of mathematics on the elementary arithmetic – never received any detailed response from Husserl" (22). Though it is true that Husserl's "inquiry into the theory of number led him into general epistemological investigations that occupied him for the remainder of his life," it should not be neglected that Husserl's *Double Lecture on the Imaginary in Mathematics* is a non-trivial attempt at dealing with the reduction of other number systems (the wholes, the rationals, the reals) and of their properties to the naturals and thus to give an answer to some of the problems left unsolved in *PdA*. From Miller 1982, too, one gets the impression that Husserl did not achieve "the philosophical project he had begun under the inspiration of Weierstrass" (9). Miller argues that "one can only conjecture about Husserl's reasoning here. Perhaps his view was simply this: Since even our most elementary number concept is largely 'symbolic', there is no intrinsic mystery regarding the introduction of other 'symbolic' concepts, such as those of negative, rational, irrational and imaginary numbers. The original or 'authentic' number concept has already been broadened to include numbers not actually given to us, so why should we not broaden it further? We are perfectly justified in taking this step . . ." What Miller does not seem to pay sufficient attention to is that in the *Double Lecture* Husserl's philosophical problem is one of a conceptual kind: formally we can extend the natural number system by dropping certain restrictions to the executability of certain operations, but we cannot expand the concept at the basis of a specific numerical field (cp. our account of Husserl´s critique of Dedekind in ch. 3 below). So Husserl's reasoning seems to be just the opposite of what Miller suggests.

[5]See *PdA* App. Abhandlung I, *Zur Logik der Zeichen* (*Semiotik*), 340–373; II, *Begriff der allgemeinen Arithmetik*, 374–379; III, *Die Arithmetik als apriorische Wissenschaft*, 380–384; V, *Zum Begriff der Operation*, 408–429; IX, *Die Frage der Aufklärung des Begriffes der "natürlichen" Zahlen, als "gegebener", "individuell bestimmter"*, 489–492; X, *Zur formalen Bestimmung einer Mannigfaltigkeit*, 493–500. See Eley, Textkritischer Anhang, 521–562. A separate treatment has to be reserved for Abhandlung V, *Zur Lehre vom Inbegriff*, 385–407, see below Appendix 3.

[6]*PoA* Foreword 7; *PdA* Vorrede 7.

[7]"A part of the psychological investigations in the present volume was already included, almost word-for-word, in my *Habilitationsschrift*, from which a booklet four galley sheets in length, titled "On the Concept of Number: Psychological Analyses" was printed in the fall of 1887 but was never made available in bookstores" (*PoA* Foreword 8; *PdA* Vorrede 8). See Miller 1982, 11; Willard, 1984, 39; Ierna 2005: "Husserl's *Habilitationsschrift* was never published and the work now known as *Über den Begriff der Zahl* is in fact just the first chapter of the *Habilitationsschrift*" (8).

number (*Anzahl*)[8] and *unity* (*Einheit*) – in so far as they are presented properly (*eigentlich vorgestellt*), i.e. intuitively given. The second part tackles the study of symbolic presentations applied to mathematics.[9]

To understand the interest in a psychological foundation of arithmetic as developed in the *Philosophy of Arithmetic*, we have to take into account both the specific historical moment at which the work was written as well as Husserl's own academic background.

In 1891 a work that aimed at laying bare the psychological foundation of arithmetic was able to arouse the interest of mathematicians and of philosophers. Since a psychologistic orientation was then dominant in philosophy, it did not appear strange at all to look for the ultimate foundation of arithmetic in this science. And then, the work fitted into the general framework of the so-called '*research on the foundations*' of mathematics, and it proposed to tackle it from a philosophical and from a mathematical point of view.[10] From the former point of view, 'research on the foundations' consisted in identifying the fundamental concepts of mathematics and examining the essence of mathematical knowledge, and it is precisely on this aspect that the first volume of Husserl's work focuses. From the mathematical point of view, research on the foundations, as it had been conducted in the second half of the nineteenth century, had given prominence to elementary arithmetic, i.e. the theory of natural numbers, as the simple and secure basis on which to found the entire edifice of mathematics. And it was precisely the issue of the "reduction" of other number systems (the wholes, the rationals, the reals) and of their properties to elementary arithmetic that was to be the object of the second volume of Husserl's work, which was never completed.[11]

---

[8]To be understood as '*finite* cardinal number' or 'natural number'. "*E. Schröder* introduced this term (*natürliche Zahl*) ... it is apparently intended to mark the distinction of the cardinal numbers (*Anzahlen*) over against the other forms of number which come into play in arithmetic: the rational and irrational, the positive, negative and imaginary numbers. Moreover the term '*Anzahl*' is not totally univocal, since it has sometimes been used to designate the concepts of numbers in series. ... Nevertheless, we have thought it most suitable in this work to adhere to the older and almost universally customary use of language" (*PdA* 114 n., *PoA* 120 n.).

[9]"In the first of its two parts, the Volume I before us deals with the questions, chiefly psychological, involved in the analysis of the concepts *multiplicity*, *unity*, and *number*, insofar as they are given to us authentically (*eigentlich*) and not through indirect symbolizations. The second part then considers the symbolic representations of multiplicity and number, and attempts to show how the fact that we are almost totally limited to symbolic concepts of numbers determines the sense and objective of number arithmetic" (*PdA* 7; *PoA* 7).

[10]See Ortiz Hill 2002, 81.

[11]As is well known, one of the traits that distinguish the mathematics of the nineteenth century from the mathematics of the preceding century, is the birth of that movement, often called the 'critical movement', characterized by the need to provide rigorous concepts and proofs for vast branches of analysis and, later on, to reconsider the foundation of mathematics. The arithmetization of analysis initiated by Weierstrass concludes with the simultaneous publication in 1872 of the foundations of the system of real numbers by Georg Cantor (1845–1918) and Richard Dedekind (1831–1916). See Kline 1972, 947–978; Casari 1973, 1 ff.

As regards Husserl's academic background, we note that while working on the *Philosophy of Arithmetic* he still was – as Brentano defined him in a letter to Stumpf of 1886, asking him to support Husserl in his attempt to obtain the status of *Privatdozent* in Halle – "a mathematician interested in philosophical questions." In fact, Husserl had studied mathematics in Berlin with mathematicians of great stature, such as Kronecker, Kummer and Weierstrass, and he had been Weierstrass' assistant, working with him until about 1883.[12] The Husserl-Archives in Leuven have the following stenographical notes (*Nachschriften*) of lectures on mathematics:

1. *Einleitung in die Theorie der analytischen Funktionen* (Weierstraß, S.S. 1878)
2. *Stenographische Nachschrift der 54 Vorlesungen über die Theorie der algebraischen Gleichungen* (Ludwig [*sic*!] Kronecker, W.S. 1878/79)
3. *Einleitung in die Theorie der elliptischen Funktionen* (Weierstraß, W.S. 1878/79)
4. *Vorlesung über die Variationsrechnung* (Weierstraß, S.S. 1879), a notebook which contains an elaboration of lectures by Weierstraß in that term, made by L. Baur. Husserl employed it to complete his elaboration of lectures by Weierstraß on the calculus of variations and mentioned it therein. The notebook has on the front page solely the mark: "Edmund Husserl 1880".
5. *Theorie der analytischen Funktionen* (Weierstraß, W.S. 1880/81).[13]

During the winter semester 1884/85 and the summer semester 1886 Husserl came under the influence of the philosophy of Franz Brentano,[14] and – as his own words testify – it was precisely in virtue of this influence that he came to dedicate himself completely to philosophy:

> In a time of growing philosophical interests and of wavering, whether I should stick with mathematics for life or dedicate myself completely to philosophy, Brentano's lectures gave the breakthrough. I attended them at first out of mere curiosity, to hear the man, who at that time was the talk of the day in Vienna, venerated and admired by some in an extreme way, by (no few) others insulted as masked Jesuit, flatterer, salesman of idle chit-chat (*Friseur*), sophist, scholastic. At the first impression I was quite affected. ... Soon I was drawn onwards and convinced by the absolutely unique clarity and the dialectic acuity of his arguments, ... by the cataleptic force of his way of developing problems, theories. Most of all from his lectures I got the conviction that gave me the courage of choosing philosophy as my life-long work... [to maintain] that philosophy, too, is an area of serious work, that it also can and hence must be treated as a rigorous science.[15]

---

[12]Cf. Schuhmann 1977, 7.

[13]Eley, *Einleitung des Herausgebers*, *PdA* xxi–xxii; K. Schuhmann 1977, 6–9; Miller 1982, 2–3; Ierna 2005, 5.

[14]"*me totum abdidi in studia philosophica duce Francisco Brentano*" (Schuhmann 1977, 13).

[15]Husserl in: Kraus 1919, 153–154.

From Weierstrass, one could say, Husserl inherits the project of founding analysis on a restricted number of simple and primitive concepts,[16] and from Brentano he inherits the method for identifying these primitive concepts, namely by describing the psychological laws that regulate their formation.[17]

As we shall see, in the *Philosophy of Arithmetic* ideas and conceptions are voiced that are shared by an important part of the Berlin mathematical school in which Husserl had been active. Especially with respect to the specific Husserlian solution for the foundation of the concept of number based on its psychological constitution, it was quite usual among the Berlin mathematicians to introduce the concept of natural number in a psychological way. During his lecture course *Introduction to the theory of analytical functions* of the summer semester 1878 (of which Husserl had taken and elaborated notes, as reported above), Weierstrass himself begins the first class with a psychological characterization of the concept of number. He introduces it (just as Husserl will do in the *Philosophy of Arithmetic*, but in a different theoretical vein[18]) as the result of a subjective activity, that of counting, which constitutes "an object of thought," i.e. the number:

> We best attain the concept of number by proceeding with the operation of counting. We consider a given aggregate of objects; among these we look for the ones that have a certain feature apprehended in the presentation by going through them sequentially; we comprehend the single objects with the feature together in an encompassing presentation, and thus a multiplicity of unities is made, and this is the number.[19]

And already in a lecture from 1874 Weierstrass explained that "a natural number is the representation of the collection of things that are the same".[20] However, these were introductory definitions that had didactical purposes and were not referred to again in the lecture course.

---

[16]"It was my great teacher Weierstrass who, through his lectures on the theory of functions, aroused in me during my years as a student an interest in a radical grounding of mathematics. I acquired an understanding for his attempts to transform analysis – which was to such a very great extent a mixture of rational thinking and irrational instinct and knack – into a rational theory. His goal was to set out its original roots, its elementary concepts and axioms, on the basis of which the whole system of analysis could be constructed and deduced by a fully rigorous, thoroughly evident method" (Schuhmann, 7). Among the interpreters who see in Weierstrass the source of Husserl's interest for "a radical grounding of mathematics" we find Willard 1984, 21–23; Miller 1982, 3 ff.; Ortiz Hill 1994a, 2 ff. & 1997b, 139 & 2004, 123–124; Ierna 2005, 3 ff.

[17]As Miller 1982 puts it "... Husserl's philosophy of arithmetic took shape as an attempt to address the non-mathematical issues to which the program of arithmetizing analysis inevitably led". (4) See also ibid., 19.

[18]Cp. Miller 1982, 4, 6, 8.

[19]Weierstrass, *Einleitung in die Theorie der analytischen Funktionen* (lecture of May 6, 1878), notes by Husserl, English transl. from Ierna 2006, 36 f. Cp. also Miller 1982, 3.

[20]Weierstrass 1966, 78.

1.2 'Many As One': The Concept of Multiplicity (or Set)

The first four chapters of the first part of the *Philosophy of Arithmetic* revolve around the analyses of the concepts[21] of *multiplicity* or *set* and *cardinal number*. Before going into the exposition of its contents, we notice that Husserl uses the terms 'multiplicity (*Vielheit*)', 'plurality (*Mehrheit*)', 'set (*Menge*)', 'collection (*Inbegriff, Sammlung*)' and 'aggregate (*Aggregat*)' essentially as stylistic variants.[22] The presence of different names for the same conceptual content is thought to express the difference between the *distributive* and *collective modes* of conceiving a whole made up of discrete objects.[23]

The problem under consideration here is that *of the unitary treatment of a multiplicity of objects, of the way we connect a multiplicity of things*, in other words, *of the way we unify a multiplicity of objects into a new object*. This is, at bottom, the main problem at the core of all set theories.[24] That Husserl understood this issue exactly in these terms is confirmed, among other things, by the following observation: "But how is this remarkable fact itself to be explained... that the same content appears to us now as 'one' and now as 'many'?"[25]

In general Husserl distinguishes with respect to a concept its extension (*Umfang*), its content (*Inhalt*), and its genesis (*Entstehung*). By 'extension' he means more or less what is also meant today, i.e. the class of objects that fall

---

[21]In his early writings Husserl "speaks of number, the concept of number, and the representation of number, in quite the same way" (Willard 1984, 26). (Hence we disagree with the following contention in Miller 1982, 22: "for the Husserl of *PoA* numbers are not presentations; they are rather concepts which are 'contained' in certain presentations".) Willard also gives a justification for Husserl's interchangeable use of the expressions "the concept of number" and "number", and of "analysis of the concept of number" and "analysis of number". This use has been largely followed in the secondary literature, and is adopted here as well. "Conceptual analysis is simultaneously an analysis of the essence of an object *insofar* as it is an object of the concept in question. ... The literature of recent philosophy contains many passages where these terms are used interchangeably, and this also occurs in Husserl" (loc. cit.).

[22]*PoA* 15; *PdA* 14. It is noteworthy that in §82 of the *Wissenschaftslehre* Bolzano defines 'collection (*Inbegriff*)' as the comprehensive union (*Zusammenfassung*) of at least two arbitrary objects (concrete or abstract) – called parts of the collection – in a whole. Furthermore, he defines the concepts of set (*Menge*), sum (*Summe*), series (*Reihe*) as suitable specializations – through the specification of the kind of connection – of the concept of '*comprehensive union of the parts in a whole*'. Here is a rough summary: A collection is a *sum* if and only if (i) it contains the parts of its parts, and (ii) it is invariant with respect to the permutations of its parts. An aggregate is a *set* if and only if (i) it does not contain the parts of its parts, and (ii) it is invariant with respect to the permutations of its parts. Finally, a collection is a *series* if and only if (i) it does not contain the parts of its parts, (ii) it is not invariant with respect to the permutations of its parts, and (iii) it has an ordering relation. Cp. Bolzano, *WL* I, §§82–85; and the detailed critical reconstruction in Krickel 1995. Cp. also Simons 1997 & Simons (ms.); Behboud 1997. While Husserl does not say so explicitly, these conceptual distinctions are at work in *PdA* – in a very similar way as Bolzano intends them.

[23]Cp. Ortiz Hill 2002, 80.

[24]Cp. Casari 2000, 107.

[25]*PoA* 162; *PdA* 155.

under the concept; by 'content' he means sometimes, traditionally, the set of distinctive features of the concept, sometimes rather the intentional correlate of the concept;[26] and by 'genesis' he means the psychological constitution of the concept. As Willard puts it, "to give the origin [genesis] of a concept ... is to describe the essential course of experiences through which one comes to posses the concept."[27] The fact that each of us arrives at the same concept is warranted here, as with Brentano, by the uniformity of the ways in which the abstract concept is constituted.[28] Thus, for each of the fundamental concepts of mathematics – multiplicity, cardinal number and unity – the logical-psychological analysis must determine the extension, the content and the genesis.

Now as regards the concept of set, Husserl's argument can be summarized as follows:

1. The concept of set is an elementary concept, so *it cannot be defined*. A definition can only be given for complex concepts, and it consists in the decomposition of the concept into its components.[29]
2. The extension of this concept must be considered as something given (*ein Gegebenes*) – for when confronted with an aggregate of any objects we are always able to decide whether it is a set or not[30] –, and it is constituted by *properly presented*, i.e. *directly intuited*, unordered concrete sets.[31]

---

[26]Cp. Casari 1991, 35–49. For a different account of Husserl's characterization of the content of a concept see Willard 1984, 27.

[27]Willard 1974, 106.

[28]For this interpretation see Casari 1997a, 553–552. By contrast, Tieszen 1990 finds the paradigm for this procedure in Kant: "on the Kantian strategy human subjects are viewed as so constituted that their fundamental cognitive processes are isomorphic" (152). Referring to this process in *Formale und Transzendental Logic* Husserl says: "I acquired a determined view of the formal and a first comprehension of its sense already in my *Philosophy of Arithmetic*. Though immature ... it was a first attempt to obtain clarity on the proper and original sense of the fundamental concepts of set- and number-theory, by falling back on the spontaneous activities of collecting and counting in which collections ("aggregates", "sets") and numbers are given... *It can be recognized a priori that each time the form of this spontaneous activities remains the same, correlatively, the form of their constructions remains the same*" (*FTL* 76, my emphasis).

[29]"Mathematicians have followed the principle of not regarding mathematical concepts as fully legitimized until they are well distinguished by means of rigorous definitions. But this principle, undoubtedly quite useful, has not infrequently and without justification been carried too far. In over-zealousness for a presumed rigor, attempts were also made to define concepts that, *because of their elemental character, are neither capable of definition nor in need of it*" (*PoA* 101; *PdA* 96 (my italics)). Simons (ms.) embraces Husserl's position when he says: "It is impossible to define the general notion of a collection in terms of anything conceptually more simple, so let us simply give some examples."

[30]"No one hesitates over whether or not we can speak of a multiplicity in the given case. This proves that the relevant concept, in spite of the difficulties in its analysis, is a completely rigorous one, and the range of its application precisely delimited. Therefore we can regard this extension as a given. ..." (*PoA* 16; *PdA* 15). Cp. Ortiz Hill 2002, 81.

[31]For the distinction between proper and improper (symbolic) presentations, see §8 of this chapter.

3. The genesis of the concept of set is due to psychological *abstraction*, starting off with an unordered set of concrete objects, that constitutes the "concrete basis (*die konkrete Grundlage*) for the abstraction."[32]

In the context of (3), "abstracting" is: leaving aside the peculiar features of the elements and considering the resulting units as merely being distinct. The concept of multiplicity is the result of this process of abstraction, where the idea is that a single act of abstraction is applied simultaneously to all the members of the set and not to each single element one by one.[33]

In his *Beiträge zur Begründung der transfiniten Mengenlehre* Georg Cantor gives an account of cardinal number, which is wholly analogous to Husserl's, except for the fact that Cantor takes the cardinal number of a set to be the result of an act of *double* abstraction both from the particular features, and from the order, of the elements of a set, whereas Husserl, as we just said, "starts off with an unordered set M and obtains the cardinal number of M as the result of a *single act of abstraction* on its members."[34] In Section 1 of *Beiträge* Cantor writes:

> We will call by the name "power" or "cardinal number" of M the general concept which, by means of our active faculty of thought, arises from the aggregate M when we make abstraction of the nature of its various elements *m* and of the order in which they are given.
>
> We denote the result of this double act of abstraction, the cardinal number or power of M, by ... $\overline{\overline{M}}$. Since every single element *m*, if we abstract from its nature, becomes a "unit", the cardinal number $\overline{\overline{M}}$ is a definite aggregate composed of units, and this number has existence in our mind as an intellectual image or projection of the given aggregate.[35]

In §72 of his *Was sind und was sollen die Zahlen?* (1888) Richard Dedekind gives a similar characterization of the numerical series:

> If in the consideration of a simply infinite system *N* set in order by a transformation ϕ we entirely neglect the special character of the elements, simply retaining their distinguishability and taking into account only the relations to one another in which they are placed by the order setting transformation ϕ, then are these elements called *natural numbers* or *ordinal numbers*, and the base-element 1 is called the *base-number* of the *number-series N*. With reference to this freeing the elements from every other content (abstraction) we are justified in calling numbers a free creation of the human mind.[36]

As Kit Fine puts it, " what these accounts have in common is a view of abstraction as the process of freeing an object of its peculiar features and a conception of number ... as the product of such a process."[37]

---

[32] *PoA* 16; *PdA* 15. Cp. Ortiz Hill 2002, 81.

[33] *PoA* 19; *PdA* 18. Cp. Fine 1998, 600 (where, however, the characterization of the number concept through abstraction is discussed with reference to Cantor and Dedekind, and not to Husserl).

[34] Fine 1998, 602. Cp. Ortiz Hill 1997b & 2004; Simons (ms.).

[35] Cantor 1895, quoted after Fine 1998, 599. Cp. Ortiz Hill 1994b, 96.

[36] Dedekind 1888, 17. Cp. Fine 1998, 600.

[37] Fine 1998, 600. Fine also shows how, once the specific ontology underlying these characterizations of the concept of number by abstraction has been determined, it is possible to obtain an

In the case of the concept now under consideration, however, the passage from the concrete multiplicities to the general concept presents some peculiar difficulties, for once one abstracts from the nature of the elements the resulting units do not yet appear to constitute a whole. So the question is: *what* can we identify as *invariant* in all the possible cases of collection of elements in a whole? Generally we distinguish two ways of making one thing out of many: making a sum out of many parts and making a class out of many members.[38] The core of Husserl's argument consists here in observing that a set is not the simple sum (*Summe*) of its elements, but is constituted also by their connection within a whole (*Ganzes*). Thus, for instance, a wood or the starry heaven are something different from the trees or the stars taken together. It is the connection of the objects in the unity of their collection – the relation of 'collective connection (*kollektive Verbindung*)' – that turns certain contents into the members of a set and constitutes the essential characteristic that is common to all possible sets as such. Hence it is this connection that can be identified as the content of the concept of multiplicity.[39] Not the single members of the set but the set itself as a whole (*als Ganzes*) constitutes the object upon which the abstraction is performed.

> It is misleading to say that the collections (*Inbegriffe*) consist merely of the particular contents. However easy it is to overlook it, there still is present in them something more than the particular contents: a 'something more' which can be noticed, and which is necessarily present in all cases where we speak of a collection or a multiplicity. This is the combination (*Verbindung*) of the particular elements into the whole.[40]

1.3 The Collective Connection (*kollektive Verbindung*)

Husserl extensively discusses the nature and the characteristics of the notion of collective connection, a notion that, according to him, clearly plays a pivotal role, not only in psychological reflections on the concept of multiplicity, but also, more generally, in relation to our entire mental life: "Every complex phenomenon ..., every higher mental and emotional activity, requires, in order to be able to arise at all, collective combination of partial phenomena."[41] A similar opinion is shared by Dedekind (among others) who writes: "If we scrutinize closely what is done in counting an aggregate or number of things, we are led to consider the ability of the mind to relate things to things, to let a thing correspond to a thing, or to represent a thing by a thing, an ability without which no thinking is possible."[42]

---

equally plausible conception – though this does not imply its correctness *tout court* – as the more familiar one of Frege-Russell on the one hand and of von Neumann-Zermelo on the other. Cp. also Ortiz Hill 1994b, 96 & 1997a, 67 & b, 141–143 & 2004, 109–114.

[38] I borrow here terminology from Lewis 1991, 3.

[39] *PoA* 18–22; *PdA* 17–21. Cp. Ortiz Hill 2002, 81–82.

[40] *PoA* 19; *PdA* 18.

[41] *PoA* 78; *PdA* 75.

[42] Dedekind 1888, III–IV.

Husserl's analysis proceeds in two phases.[43] The first, contained in Chapter II,[44] consists in the critical examination of a series of theories that propose characterizations of the notion of collective connection that differ from the theory he will develop in the second phase of the analysis (Chapter III)[45] and that consequently give a different explanation of the origin of the concepts of 'multiplicity' and 'number'. Schematically put, they treat the collective connection as: (i) the simultaneous presence of the contents of the aggregate in consciousness; (ii) the temporal succession of the contents of the aggregate in consciousness; (iii) the intuitive form of time; (iv) the intuitive form of space; (v) the relation of identity of every content with itself; (vi) the relation of difference of every content from all others. The positive result of Hussel's critique of these theories is that the whole business of making many into one[46] is the result of a psychical act of a specific kind that picks up certain contents and unites them collectively. Where there are many things, then there is one thing[47] when "a unitary interest – and, simultaneously with and in it (*in und mit ihm*), a unitary noticing – distinctly picks out and encompasses various contents".[48] In a set, apart from the single objects, we do not find anything except the fact that we 'think them together'. The collective connection, though treated as a relation, is actually a *psychical act*. What justifies speaking of a relation also in this case is the *homogeneity of function* that it shares with the primary relations: that of connecting objects that are "unconnected" by themselves.[49]

1.4 The Concept of Cardinal Number (*Anzahl*)

Husserl also distinguishes with respect to the concept of cardinal number extension, content and origin. The *origin* is essentially analogous to that of the concept of multiplicity: the concept of cardinal number is an elementary concept. It is not possible to define it because it is logically simple. It is generated by (psychological)

---

[43]Willard 1984, 30 ff. provides a detailed account of this notion, in particular tracing it back to earlier sources like Lotze's account of the psychological origins of representations of relations: "In a chapter to which Husserl makes explicit reference Lotze presents his general view of how relations come before consciousness in activities of 'higher order' ... the activity of representing a relation is called 'higher' by him in a sense that precisely coincides with what Husserl later meant by the terms 'founded' and 'higher order' as applied to *acts* ... of consciousness. It is ... 'higher in that determinate sense in which the higher has the lower for its necessary presupposition' ..." (30). The interest of this issue is mainly historical, so we shall not engage with it here.

[44]*PoA* 23–65; *PdA* 22–63.

[45]*PoA* 67–79; *PdA* 64–76.

[46]I borrow this terminology from Lewis 1981, 6.

[47]Terminology again from Lewis 1981, 6.

[48]*PoA* 77; *PdA* 74.

[49]"There is *de facto* so much in common between the primary relation and the psychical relation, as to their essential Moment (*Hauptmoment*), that I fail to see why a common term would not be justified here" (*PoA* 76, n. 11; *PdA* 73, n. 1).

abstraction – a process that, so to speak, releases an object from its specific features. Its *termini a quo* are the same concrete phenomena that generate the concept of multiplicity and set, i.e. an unordered collection of arbitrary objects. And as in the other case, the process consists in disregarding the particular features of the connected members and in considering each of them only as a featureless 'something' or 'one', united with the others by way of the collective connection.[50] However, it is important that each element is still different (*verschieden*) from all the others and remains so after the abstraction.[51] 'Abstracting from the contents' does not imply that they 'disappear (*verschwinden*)', – 'abstracting from something' is rather, according to Husserl's unfortunate definition, "not paying any special attention to it (*darauf nicht besonders merken*)".[52] Here Frege's ironical comments in his *Review* of the *Philosophy of Arithmetic* certainly have a point:

> Inattention is an extremely effective logical faculty; whence, presumably, the absentmindedness of scholars. For example, suppose that in front of us there are sitting side by side a black and a white cat. We stop attending to their colour: they become colourless, but are still sitting side by side. We stop attending to their posture: they are no longer sitting (without, however, having assumed a different posture), but each one is still at its place. We stop attending to their location: they are without location, but still remain quite distinct. In this way, perhaps we obtain from each one of them a general concept of a cat. By continued application of this procedure, we obtain from each object a more and more bloodless phantom. Finally we thus obtain from each object a Something wholly deprived of content; but the Something obtained from one object is different from the something obtained from another object – though it is not easy to see how.[53]

The final question marks one of the three main difficulties that Frege sees for a naïve characterization of the essence of number (naïve because it is a theoretical use of a commonsensical way of thinking).[54] This critique can be found already in his

[50]Cp. Ortiz Hill 2004, 126.

[51]In *On the Concept of Number* Husserl characterizes the abstractive process that yields the concepts of set and number the same way: "It is easy to characterize the abstraction which must be exercised upon a concretely given *Vielheit* in order to attain the number concepts under which it falls. One considers each of the particular objects merely insofar as it is a something or a one herewith fixing the collective combination; and in this manner there is obtained the corresponding general *Vielheitsform*, one and one ... and one, with which a number is associated. In this process there is total abstraction form the specific characteristics of the particular objects ... To abstract from something merely means to pay no special attention to it. Thus in our case at hand, no special interest is directed upon the particularities of the content in the separated individuals" (Husserl 1887, 116–117). Quoted after Ortiz Hill 2002, 82.

[52]*PoA* 83; *PdA* 79. Cp. Ortiz Hill 1994b, 96–98.

[53]Frege 1894, 181. Frege has developed this criticism already 1884 in his *Grundlagen*, §§29–44, and in his hilarious 1899 he directs it against a contribution to the *Enzyclopädie der mathematischen Wissenschaften* by a *Gymnasialprofessor* in Hamburg. On Frege against psychological abstraction see esp. Dummett 1991a, 49–50 & 1991b, Chapters 8, 12, and pp. 167–168 where psychological abstraction is carefully distinguished from logical abstraction as used, for example, in Frege's contextual definition of the direction-operator. Cp. also Ortiz Hill 1994b, 96–98 & 2004, 114; Tieszen 1994, 318.

[54]Cp. Ortiz Hill 1992b, 98 & 1997, 66.

Grundlagen (§§29–44) as an objection against the definition of numbers as sets of units, and it is widely accepted in analytical circles as the main objection to all theories that propose a definition of natural number by (psychological) abstraction.[55] On Frege's account, the units of which we speak can be neither the things that are counted, otherwise there would be as many numbers two as there are groups of two things; nor can they be the things after abstracting from their properties. Indeed, (1), if we abstract from all the properties of the things, we have by the Leibnizian principle of the *identity of indiscernibles* just one single unit and hence no sets of units; and, (2), if we do not abstract from all the properties, but only from those that distinguish the objects of a certain group, we will not arrive at a number – or at a numerical concept – but simply at a general concept: e.g., starting from the things *a* and *b*, we will not get to 2, but to the concept 'thing that has the properties common to *a* and *b*'.[56]

Coming back to Husserl, the abstract form 'one and one and one . . .', generated by abstraction, constitutes, according to him, a specific type of structure. This is the content (*Inhalt*) of the concept of set as well as of the concept of cardinal number, only that in the latter case we find associated with that abstract form a numeral (*Zahlwort*). The cardinal number is hence conceived, in a Bolzanian way, as a property that characterizes a set of objects, i.e. as the measure for the width of a set. It worth recalling here that Cantor wrote in a letter to Giuseppe Peano: "I conceive of numbers as 'forms' or 'species' (general concepts) of sets."[57] Husserl's view must be that numbers are certain *non-distributive* properties of sets.[58] "'Non-distributive' means that just because the multitude has the property, it does not follow that its parts, or submultitudes, in particular its members, have it."[59]

The *extension* of the concept of cardinal number is constituted by the numbers of the series. Cardinal numbers answer to the question *How many?* They represent the exact *determination* of the elements of a concrete set or of a set while abstracting from its elements (set of units). The number series is founded on the possibility to distinguish and classify all the possible sets of units, and to order them by the order relation. Between the general concept of 'cardinal number' and the numbers of the series there obtains a genus-species relation: the single numbers are the *species infimae* of the genus 'cardinal number'.

---

[55]Cf. Fine 1998, 604–605.

[56]In his *Review* Frege ironically concludes that, from what Husserl writes, to obtain the concept of number, one must exercise abstraction only up to a certain point, i.e. the point at which the members of the set no longer have any specific properties, but are nevertheless still distinct. "Number-abstraction simply has the wonderful and very fruitful property of making things absolutely the same as one another without altering them. Something like this is possible only in the psychological wash-tub" (Frege 1894, 188). Cp. Ortiz Hill, 1994b, 97.

[57]Cantor 1991, 365.

[58]I borrow here terminology from Simons 2007, 233.

[59]Loc. cit.

1.5 Chapters VI and VII of the *Philosophy of Arithmetic*

The reflections and discussions Husserl develops in Chapters VI and VII are, in many respects, among the most interesting of the first part of the *Philosophy of Arithmetic*. Chapter VI, "*The Definition of Equinumerosity through the Concept of Bijection*," opens with a discussion of the definition of the general concept of equality (*Gleichheit*), proceeding then to focus on a critical examination of the definition of equinumerosity (*Gleichzahligkeit*) through bijection, i.e. one-to-one correspondence (*gegenseitige Zuordnung*), and on an analysis of the essence (*Wesen*) of one-to-one correspondence as a relation. Chapter VII, "*Numerical Definitions Through Equivalence*," contains both a detailed critique of the definition of the concept of natural number *by logical abstraction* in Frege's *Grundlagen der Arithmetik* (1884) and a general critique of the practice of defining a concept by defining its extension.[60] The latter is in fact the true focal point of Husserl's complex argumentation: it occurs first in his rejection of theories that define the concept of *having the same number* through one-to-one correspondence, and it reappears in his criticism of the definition of the concept of natural number through equality (i.e. equality as standing in the relation of one-to-one correspondence).

At the end of these two chapters, Husserl thinks he has adequately justified the thesis that the concept of equivalence does not occur in any way in the constitution or the definition of the concept of *Anzahl*.

Husserl's critique of Frege in Chapter VI opens with a consideration that we already know: "As soon as we come upon the ultimate, elementary concepts, all defining comes to an end."[61] All that we can ask of the exposition (*Darstellung*) of an elementary concept is that it enables us to reproduce in ourselves those psychical processes that are necessary for the constitution of the concept.[62] Since they are *elementary*, neither the broad concept of *equality* nor the narrower concept of *equality of two sets with respect to their number* (i.e. *equinumerosity*) can be defined.

Frege placed the Leibnizian definition[63] of *equality* as *substitutivity salva veritate* ("*eadem sunt quorum unum potest substitui alteri salva veritate*"), which was also used by Grassmann,[64] at the foundation of his definition of the concept of number.

---

[60]Cp. Ortiz Hill 2002, 95–96 & Tieszen 1994, 320.

[61]*PoA* 125; *PdA* 119. Cp. Tieszen 1990, 152 & 1994, 320.

[62]Cf. *PoA* 125; *PdA* 119.

[63]*Definitio prima*, in Leibniz 1687.

[64]Grassmann 1844 represents an important moment in the history of the development of mathematics from a theory of magnitudes to a theory of forms. The substitution of the concept of magnitude with that of form in the definition of mathematics is not, however, to be considered an anticipation of the modern conception of mathematics as theory of structures, but, rather, in the sense of a conception of the objects of mathematics as forms of thought, objects of thought. On Grassmann also cp. Webb 1980, 44 ff. On Grassmann's influence on Husserl see e.g. Hartimo 2007, 292 ff.

This attempt at a definition has an odd feature which neither Frege nor Husserl comments upon. Take "Hesperus is Phosphorus". If the substitution is to take place in a sentence, what is exchanged can hardly be a planet: a sentence does not contain any heavenly bodies, and then, what could it possibly mean to replace a planet by itself? Leibniz's "*unum*" and "*aliud*" make sense, however, as soon as we take him to mean (but unfortunately failing to say) that the singular terms that flank the identity operator can be exchanged *salva veritate*. But alas, though this makes sense it is not true, and Leibniz himself came to recognize this.[65]

According to Husserl, Leibniz's attempt at a definition is misguided in any case, and furthermore it fails for the following reasons. First, it defines identity instead of equality: "So long as there is a remainder of difference, there will be judgments in which the things under consideration cannot be substituted *salva veritate*."[66] Secondly, the fact that two contents can be substituted *salva veritate* is not the *reason* of their equality: on the contrary, their equality is the reason of their substitutivity *salva veritate*.[67] And thirdly it does not provide us with a *criterion for recognizing* the equality: in fact, proving the substitutivity *salva veritate* of two contents a and b leads back to the evaluation of an infinite number of equalities – the equality of the truth value of $A(a)$ with that of $A(b)$, for all possible judgments A – and so on *ad infinitum*.

Now Husserl's analysis moves to a specific notion of equality, the *equality of two multiplicities with respect to their number*, or *equinumerosity*. To find some firm ground to approach the question, Husserl takes as reference point the following definition by Stolz: "Two multiplicities are said to be *equal* [or, more correctly: *equally many*, *equinumerous*] to each other if each thing of the first can be correlated with one thing of the second, and none of these remain unconnected."[68]

There is more than one reason why Husserl thinks this definition is not a "good" definition of equinumerosity. First of all, it is circular: the concept of (numerical) equality is defined by implicitly invoking the concepts 'more' and 'less', which presuppose it.[69] Furthermore, it cannot be considered even a nominal definition of the concept of equinumerosity, as *definiens* and *definiendum* ('being two equal

[65]Leibniz's and Frege's reasons for restricting the substitutivity claim are explained and discussed in Künne 2009, Ch. 1, §5. Here one also finds additional reasons that Leibniz and Frege did not yet take into account.

[66]*PoA* 102; *PdA* 97.

[67]Cp. Ortiz Hill 1994a, 5–11.

[68]*PoA* 103; *PdA* 98. Otto Stolz (1842–1905) was an influential Austrian mathematician (professor at the University of Innsbruck since 1872 until his death), with major interests in algebraic geometry and analysis. Husserl's quotation is taken from Stolz 1885. Incidentally, Stolz is the first mathematician who wrote a paper on Bolzano: see Stolz 1881.

[69]"We can see that the presentation of 'more' and 'less' is already included in the definition of equality, while these themselves ... cannot be conceived without presentations of equality. When we say that the bijection must not leave any element unconnected, then this is just a different way of saying that on neither side there can be an element more or less. Thus the circularity is obvious" (*PoA* 103–104; *PdA* 98–99).

multiplicities' and 'being two multiplicities in one-to-one correspondence') are not *conceptually* equivalent.[70]

To assess the equality of two sets we must first know in what respect the two sets are to be compared, because the same objects can be judged to be equal or unequal depending on the characteristics on which we focus our interest at a given moment.[71] If the interest is directed at the *number of elements* contained in each set, then there are two possible ways of proceeding in the comparison. The *first* consists in trying to put the elements of the two sets into one-to-one correspondence, and in verifying that no element remains unconnected. The *second* consists simply in *counting* the elements of the two sets and verifying in this way whether or not they have the same number.

The first method can be used when we just want to assess the mere equinumerosity, without wanting to know the precise number of elements of each set; whereas the second method is used when we actually want to know the cardinal number by counting the elements of the two sets *in the symbolic sense*. The evident advantages of the second method, that make it preferable to the first, are essentially the following three: (i) it is a completely *mechanical* process, which can be executed without thinking about the concepts involved; (ii) it is efficient and secure; and (iii) it enables us not only to compare the two sets, but also to obtain the cardinal number associated with each of them.[72]

This should serve to clarify once and for all the sense and role that Husserl assigns to one-to-one correspondence, and the intrinsic limitation that he finds in the definition of 'equinumerosity' under consideration: relative to sets with a *finite* number of elements, one-to-one correspondence '*warrants* (*verbürgt*)' equinumerosity, but it is not *what determines* equinumerosity. "The possibility of bijection between two sets *is* not [the reason of] their equinumerosity, but only *warrants it*."[73]

So the reason why two sets have the same cardinality is *not* the fact that they can be put into one-to-one correspondence; on the contrary, the correspondence is possible *only if* the two sets have the same cardinality. Trying to put them into one-to-one correspondence clearly is an operation that is always meaningful and that can have some *practical* value, especially when the number of the elements is high (but still *finite*). "It may well happen that, in order to verify *in concreto* the equality of two sets with respect to their multiplicities [sc.: with respect to the number of their elements], we place pairs of elements alongside one another or

---

[70]Cp. Tieszen 1990, 153.

[71]"If there is equality in the (internal or external) characteristics that at that moment constitute the focal point (*Mittelpunkt*) of our interest" (*PoA* 105; *PdA* 100).

[72]"What is simpler than comparing the two multiplicities with respect to their number by counting them both in the symbolic sense? Hence, we obtain not only the assurance of the equality (or inequality) of the numbers, but also *these numbers themselves*. That the mechanical process of counting, already for sets with a relatively low number, proceeds in an incomparably faster and more certain way than that apparently simple process of bijection, surely does not need a demonstration" (*PoA* 109–110; *PdA* 104–105).

[73]*PoA* 110; *PdA* 105.

connect them in some other way; but *neither can we consider this operation necessary everywhere*, nor, where this happens, *the essence of the act of comparison resides only in this*."[74]

In other words: 'to have the same cardinality' and 'to be in one-to-one correspondence' are not concepts with *the same content*, but only concepts with the *same extension*. The definition above, then, cannot be a nominal definition of the concept 'equality of two multiplicities with respect to their number'. If giving a definition means univocally fixing a concept, then Husserl's objection – put forward again in his critique of the "theory of equivalence" – is that *a concept cannot be defined by defining its extension*. The definition only formulates a *logically necessary and sufficient criterion* (*Kriterium*) for establishing the existence of equinumerosity.[75]

As regards the essence of the one-to-one correspondence as a relation, Husserl takes it to be a special case of the relation of *collective connection* (*kollektive Verbindung*), limited to pairs of elements. Frege thinks that two sets can be put in bijection by *any* relation ϕ (taking 'a has relation ϕ to b' and 'a is in bijection with b' to be conceptually equivalent). For Husserl, on the contrary, it is possible to use any relation ϕ to obtain the correspondence (like 'put the elements side by side', 'order the elements in pairs', etc.) exactly because the pairs of elements are previously collectively connected in our thought. It is the collective connection that makes the correspondence possible, while to any primary relation can be attributed the practical value that it makes easy to see the equinumerosity of the two sets. Hence, it is not the relation ϕ that establishes the correspondence, but the *collective connection*, i.e. the act of thinking together, in ordered pairs, all the elements of the sets that we want to compare.

From Husserl's reflections up to now clearly emerges that all the discussions are aimed at the characterization of the field of *finite* cardinal numbers (even if Husserl never explicitly says so). In his *Philosophy of Arithmetic*, Husserl says that through symbolic calculation we can count, at least in principle, any finite set and find its corresponding cardinal number, no matter how big. However, in his study "*Zur Lehre vom Inbegriff*"[76] he declares that in the first part of his book he was thinking only of *proper* numerical presentations. For these it is not necessary to put all their elements into one-to-one correspondence if one wants to assess the equinumerosity of two sets. He also admits that, as far as *symbolical* numerical presentations are concerned, there is no unitary *a priori* principle to establish whether it is also possible to classify infinite sets according to 'more' and 'less', and that to obtain such a classification, bijection is a necessary condition.[77] Again, in the same study, he recognizes the error of beginning the systematical treatment of arithmetic with

[74] *PoA* 104; *PdA* 99.

[75] "Although it is not necessary to make the comparison through bijection, *we are able to do it in all cases...* In this consists, accordingly, the only meaningful and useful application of the 'definition'" (*PoA* 110; *PdA* 105; my italics).

[76] According to the editor of *PdA* this study is from 1891. However, as we will show later, this dating is clearly mistaken.

[77] *PoA* 359–383; *PdA*, App. 385–407.

the series of natural numbers: in this way, indeed, we tacitly presuppose finiteness without defining this concept and without using it systematically:

> Tacitly, use is made of a fundamental presupposition, that one would limit oneself to *finite* numbers alone, construct an arithmetic only for them, and then also only for them all the basic principles (*Grundsätze*), that are to be established, hold true. A rigorous system of arithmetic must therefore begin with the precise distinction of numbers into finite and infinite, and then on the basis of this distinction, provide proof of the complete classification of the field of finite numbers by means of the series of natural numbers[78]

Moreover, he emphasizes that it is exactly this important distinction between finite and infinite cardinals that is neglected in the first volume of the *Philosophy of Arithmetic*.

As we already anticipated, Chapter VII is dedicated to the critique of the definition of the concept of natural number (*Anzahl*) through the concept of one-to-one correspondence. Instead of taking in consideration one by one the various theories that take this route[79] (including Frege's, which is discussed in a separate section),[80] Husserl deems it more useful to introduce and then discuss a sketch of a theory, the "theory of equivalence," that, according to him, is capable of encompassing and synthesizing all the essential aspects of the issue. It can be presented schematically (with a minimum of formalism) as follows:

1. Given two sets A and B, we say that:
   - A and B are *equivalent* (Husserl prefers this instead of 'equinumerous') when there is a one-to-one correspondence ϕ between A and B:

$$\forall x(x \in A \to \exists y(y \in B \land \phi xy))$$
$$\forall x(x \in B \to \exists y(y \in A \land \phi yx))$$
$$\forall xyz(\phi xy \land \phi xz \to y = z)$$
$$\forall xyz(\phi xz \land \phi yz \to x = y)$$

   - A is less than B when A is equivalent to a proper part of B.
   - A is more than B when B is equivalent to a proper part of A.[81]

The definitions of the relations of 'equivalence (*Gleichviel*)', 'more (*Mehr*)' and 'less (*Weniger*)' are wholly independent from the concept of cardinal number, so that, in order to decide whether or not they obtain with respect to two given sets, not

[78] *PoA* 374; *PdA*, App. 399.

[79] Among which Husserl explicitly mentions Stolz 1885.

[80] Cp. Tieszen 1994, 319.

[81] Here we still have the assumption that A and B would be *finite* sets. Moreover, Husserl's tacit assumption is that all sets under consideration contain at least two distinct elements. On this latter point see Appendix 3.

only is it not necessary to count the elements of the two sets, but it is not even necessary to know what counting is.

2. To each concrete set M we can associate the class K of all the sets N, "given or thinkable (*gegebene oder denkbare*)", that are equivalent to M, which we call "the *class of* M (*die zu M gehörige Mengenklasse*)".
3. The system of classes is a partition of all the sets:

   - No set can belong to different classes
   - Every set falls in one (and, for (a), only one) class

Consequently, every class is univocally determined by each of its members; each set of each class K can serve as a foundation (*Fundament*) for the construction of K and as a representative (*Repräsentant*) of K. From a single concrete set, Husserl says, we generate the entire corresponding equivalence class by indefinitely varying the kind of elements of the set.

Up to this point, the "theory of equivalence" looks like an analogue of what today we call 'making the quotient of a collection *modulo* a certain equivalence relation'. The system of what Husserl calls 'classes' is the quotient of the collection of all the sets (tacitly presupposed: finite sets) *modulo* that specific reflexive, symmetrical and transitive relation (equivalence relation) that is the relation of equinumerosity.

4. On the system of classes we can furthermore impose an *ordering principle*: given a class K and a set M such that $M \in K$, we consider a set M' obtained by eliminating from M any single one of its elements, and then we construct the equivalence class K' of M', which is the preceding class (*nächstniedrigere*) of K.[82] Analogously we can construct the succeeding class (*nächsthöhere*) of K, adding to the initial set $M \in K$ a new element and moving on to the equivalence class of the set we obtain. Naturally – even though Husserl does not explicitly say it – this "ordering principle" is only meaningful for finite sets: only with the added hypothesis that the sets under consideration are finite, can we produce a univocal ordering of all the classes in a series.
5. At this point we can proceed to the definition of the concept of number. Each class encompasses the totality of thinkable sets with a certain number of elements. Each set M of a given class K hence has the same cardinal number. All sets M belonging to a certain class K must have a certain shared quality (*Beschaffenheit*) that distinguishes them from the members of all other classes. This shared quality is precisely the fact that they all belong to the same class, i.e. that they are all pairwise in a relation of equivalence. Husserl says, then, that we need a suitable notation to express this shared quality in such a way as to "reflect (*widerspiegeln*)" the system of classes in its natural ordering. Because a class can be represented by any member, we choose as *representatives* of the classes,

[82]Obviously, there is here the tacit assumption that some (equivalently: every) set M in K has at least three elements.

sets of bars (*Striche*): 11, 111, 1111...; or rather, to avoid any confusion with the notation of numbers in the decimal system, sets of bars combined with the '+' sign: 1 + 1, 1 + 1 + 1, 1 + 1 + 1 + 1 ...[83] Then we identify these sets of bars with the natural numbers. "These sets of bars are the natural numbers, because as representatives of the classes they are also representatives of the number concepts."[84] Counting a concrete set M means to put it in one-to-one correspondence with a set of bars, so that we can subsume it under the class that it belongs to. "The numbers constitute an ordered series, corresponding to the series of the classes."[85]

It is important to observe that, in the "theory of equivalence," the numbers are not identified *tout court* – à la Frege – with equivalence classes, but with *canonical* representatives of the classes, i.e. with certain specific sets.

Husserl's critique of the "theory of equivalence" consists essentially in arguing that it is based on a wrong interpretation of the concept of one-to-one correspondence, which is nothing but a simple *criterion* to establish equinumerosity. The discussion developed in Chapter VI disputed that equality among sets would be the sufficient reason for the fact that two sets have the same number of elements, because "what equivalent sets have in common is not merely the 'equinumerosity' or, more clearly, equivalence, but rather the same cardinal number in the true and proper sense of the word."[86] This argument is now used to criticize the definition of the concept of number which is at the base of the "theory of equivalence."

The theory defines two sets as having the same cardinality if and only if they are equivalent, and considers this as a nominal definition. However, 'to have the same cardinality' and 'to be equivalent' are concepts with the same extension, but not with the same *content*. To take them as being the same concept leads to considering equality as *source* of the concept of number and to conclude that all sets M belonging to a given class K have nothing in common except equinumerosity. Consequently, according to this theory, belonging to a class becomes essential for the concept of number, while establishing the number of elements of a certain set M means inserting this set in a certain class of equivalent sets.

---

[83]On the fact that Husserl starts counting from the number 2, see below. Furthermore, note the similarities with Schröder's idea that the number *represents* (not *means*) the counted objects: "To obtain a sign, capable of expressing how many of those unities are present, we direct our attention step by step to each of the units under consideration, and we represent them with a small bar (*Strich*): 1 (i.e. with a numeral 1, a one). Then we put them in line, and to avoid conflating them into a number 111, they are connected with the + sign. Hence we obtain a number of the type: 1 + 1 + 1 + 1 + 1" (Schröder 1873). Cp. the following passage from the *Philosophy of Arithmetic*: "We take the sets 11, 111, 1111, ... , obtained by repeating the bar (*Strich*) '1' or the sound complex 'one', or (to avoid any confusion with certain composite signs of the decimal number system) the sets 1 + 1, 1 + 1 + 1, 1 + 1 + 1 + 1, ... , as representatives of the classes and name them according to the series 2, 3, 4 ..." (*PoA* 110; *PdA* 105).

[84]*PoA* 119; *PdA* 113.

[85]Loc. cit.

[86]*PoA* 122; *PdA* 116.

The core of Husserl's argument is that a numerical assertion does not at all express a certain *relation* of equivalence between a certain set and all possible equivalent sets.

> If cardinal numbers are defined as those relational concepts founded on equivalence, then surely every numerical assertion (*Zahlaussage*), instead of being directed upon the concretely given set as such, would always be directed upon its relations to other sets. To ascribe a determinate number to this set would mean to classify it within a determinate group of equivalent sets, but this is absolutely not the sense of a numerical assertion. Let us consider a specific example. Do we call a set of nuts lying before us 'four' because it belongs to a certain infinite class of sets that can be mutually put in bijection? Surely no one has ever thought of such a thing in this context, and we would be hard put to find any practical occasion whatever that would make it of interest. What does in truth interest us is the fact that here is a nut and a nut and a nut and a nut. We immediately adapt his unsuitable and cumbersome presentation ... to a form that is more convenient for thought and speech, by considering it by way of the general form of sets 'one and one and one and one', which has the name 'four'.[87]

In the same way, natural numbers cannot be represented by sets of bars simply because of the equality of their representatives with a class of equinumerous sets, but because we abstract from the specific nature of the members of each set, regarding them merely as indeterminate 'somethings (*Etwas*)' and designating them with the sign "1" or with a bar. "We certainly do not ascribe the number four to a set of nuts ... just because this set can be "mapped" (*abgebildet*) to '1111', and each individual nut to '1'! ... Therefore, for each content, the bar '1' can only designate that it is a 'something', and, accordingly, cardinal number is 'something and something ... etc.'."[88] The concept of set and the concept of cardinal number – this is Husserl's position – are obtained by abstraction from a randomly chosen unordered concrete set, disregarding the specific nature of its elements. *The reason we form such concepts is their practical usefulness for rational thought and language.*

In an interesting note, Husserl briefly mentions Cantor's ideas about the issue at hand. He observes that only in an initial phase of his reflections Cantor was apparently moving in the direction of the "theory of equivalence" – as is testified by the following passage from the *Grundlagen einer Allgemeinen Mannigfaltigkeitslehre* (25): "To each well-defined set ... belongs a determinate power [*Mächtigkeit*, Cantor here uses 'power' for 'cardinality'], the same power being ascribed to two sets if they can be reciprocally put in bijection element by element". But in all his later publications Cantor takes a completely different position.[89]

In fact, already in a letter to Lasswitz (from 15 February 1884, published in the *Mitteilungen zur Lehre vom Transfiniten* in 26), Cantor radically changes the definition we quoted above, giving a definition of cardinality or power of a set that is very close to the Husserlian definition in the *Philosophy of Arithmetic*:

---

[87] *PoA* 122; *PdA* 116. Cp. Tieszen 1990, 156–157.

[88] *PoA* 123; *PdA* 117.

[89] "Nevertheless, this mathematical genius in no way belongs to the tendency the criticized above, as is apparent from all his later publications." *PoA* 121, note 3; *PdA* 115, note 2.

> By *power* or *cardinal number* of a set M, we mean the universal or general concept that we obtain when we abstract from the quality of the elements of a set as well as from the relations that exist among them or with other objects and hence also, specifically, from the order that might exist among the elements, and we reflect only on what is common to all the sets that are equivalent to M.[90]

The similarity of Cantor's and Husserl's position emerges, even more clearly, from another passage of the *Mitteilungen* that Husserl quotes in the above mentioned note: "For the formation of the general concept 'five' there is required only one set [...] to which that cardinal number belongs." This method of defining the concept of set by abstracting from the nature of the elements (and in Cantor explicitly also from their order), and the refusal to use one-to-one correspondence to define equinumerosity are common to the two authors.[91]

1.6 Husserl and Frege's Theory

Husserl appends two notes to his general critique of the "theory of Equivalence" that discuss Frege's and Kerry's attempts to define the concept of natural number.[92] We will only discuss the former here and we will mention, schematically, the main points of divergence between Husserl and Frege regarding their conception of arithmetic.

First of all, we have to point out that the Fregean definition of the concept of natural number does not wholly fall under the theoretical model described by the "theory of equivalence." In this theory a number is the shared feature of all the sets belonging to a determinate equivalence class K, but it ends up being identified with a canonical representative of the class. For Frege, however, numbers are the equivalence classes themselves.[93]

Husserl repeatedly stresses how deeply Frege's conception of arithmetic and the goals of his research diverge from his own: Frege wants to found arithmetic on logic, by formally defining arithmetical concepts and formally deriving arithmetical truths in pure logic, and to exclude any "intrusion" of psychology into arithmetic.[94]

---

[90]Cantor 1887–1888. As Ortiz Hill 1994a states, "enough kinship is apparent between Husserl's and Cantor's work to have prompted scholars to speak of the influence Husserl may have had on Cantor's work". Indeed such an influence has been suspected e.g. by Cavaillès 1962 and Casari 1991. We do not agree, for Cantor's letter to Lasswitz dates from 1884, whereas Husserl's first explicit definition of sets and cardinal numbers by *Cantorian* abstraction dates from 1887. One should rather assume an influence in the opposite direction.

[91]Cp. Ortiz Hill 1994b, 103 & 1997b, 143 & 2004, 112–114.

[92]On Kerry cp. Picardi 1994.

[93]Extensionally speaking. Actually, for Frege, numbers are extensions of second-level concepts, i.e. if F is a concept, the *number of F* is the extension of the concept 'concept equinumerous to *F*.'

[94]"A foundation of arithmetic on a series of formal definitions out of which all the theorems of that science could be deduced purely syllogistically is Frege's ideal" (*PoA* 124; *PdA* 118). Cp. Ortiz Hill 1994b, 101–104; Tieszen 1994, 318.

For Husserl as well as for Frege, the natural numbers are those numbers that answer the question "How many?" However, for Frege numbers are not (i) properties of spatio-temporal objects, (ii) mental representations, or (iii) sets of units. In particular, point (ii) implicitly represents a sharp critique to the kind of psychologism that, according to Frege, underlies all of Husserl's arguments in the *Philosophy of Arithmetic*. If natural numbers were mental representations – Frege says – there would be as many numbers 2 as there are minds that think "2." Furthermore, numbers would be psychophysical facts and hence, together with their properties, subject to natural evolution and change. Finally, the existence of those numbers that exceed our subjective presentational capacities would be in doubt, and hence also the existence of infinite numbers.[95] For Frege, psychology must be completely excluded from arithmetic: in particular, (i) numbers are not subjective entities, they are objective, i.e. 'not subjective' *par excellence*, independent from our sensation, intuition, presentation, imagination and mental pictures. Moreover, (ii) descriptions of internal processes that precede the formulation of numerical judgments do not have any relevance to arithmetic.

Husserl rightly says that in the *Grundlagen*, Frege laments the presence of psychological argumentations in mathematical treatises: "when we feel the need to give a definition without being able to, we want at least to describe how to reach the object or concept under consideration."[96] We have seen how, in the *Philosophy of Arithmetic*, the introduction of fundamental concepts of set theory and the theory of cardinal numbers, in addition to the characterization of the very nature of

[95]Frege, *Grundlagen der Arithmetik*, §27. By way of an example, in the first volume of the *Grundgesetze der Arithmetik* (1893), criticizing the fact that Dedekind's notion of 'system' seems to admit a psychological foundation, even though Dedekind was not interested to give such a foundation, Frege writes: "This holds especially of what mathematicians like to call a 'set.' Dedekind [*Was sind und was sollen die Zahlen*] uses the word 'system' with much the same intention. But despite the explanation that appeared in my *Foundations* four years earlier, he lacks any clear insight into the heart of the matter, though he sometimes comes close to it, as when he says (p. 2): 'Such a system S ... is completely determined if, for every thing, it is determined whether it is an element of S or not. The system S is therefore the same (*dasselbe*) as the system T, in symbols S = T, if every element of S is also an element of T, and every element of T is also an element of S.' In other passages, however, he goes astray, e.g., in the following (1–2): 'it very often happens that different things *a*, *b*, *c* ... regarded for some reason from a common point of view, are put together (*zusammengestellt*) in the mind, and it is then said that they form a system S.' A hint of the truth is indeed contained in talk of the 'common point of view'; but 'regarding', 'putting together in the mind' is no objective characteristic. I ask: in whose mind? If they are put together in one mind, but not in another, do they then form a system? What may be put together in my mind must certainly be in my mind. Do things outside me, then, not form systems? Is the system a subjective construction in the individual mind? Is the constellation Orion therefore a system? And what are its elements? The stars, the molecules or the atoms?" (Frege 1893, 1–4, transl. Beaney 1997, 208–211).

[96]Op. cit., 218.

number, happens precisely by describing the *psychological process* through which we reach these concepts.[97]

Let us now turn to the Fregean definition of the concept of natural number. Husserl's critique proceeds from a detailed exposition of the content of §§63–69[98] of the *Foundations of Arithmetic*, in which Frege gives a clear exposition of the famous *method of definition by logical abstraction*, on which he bases his definition of the concept of natural number.

"The relationship of equality – Frege says – does not hold only amongst numbers... We intend ..., by means of the concept of equality, taken as already known, to obtain that which is to be regarded as being equal."[99] The Fregean method, in its generality, can be summarized as follows. Let C be a concept, for which it makes sense to speak of the objects 'the C of x', 'the C of y', etc., where x, y, ... are elements of an appropriate domain D – think e.g., of concepts like 'extension' (of a concept), 'course of values' (of a function), 'direction' (of a line) or just 'number' (of a concept). Suppose an explicit definition of C is not easily available, but that there exists a binary relation R (on the domain D) that does not presuppose C, and satisfies, for all x and y in D:

$$R(x, y) \text{ iff } x \text{ has the same } C \text{ as } y.$$

By the latter condition, R is an equivalence relation. Frege would then define:

- The C of x $=_{df}$ the extension of the concept 'thing (in D) which stands in R to x'
- C $=_{df}$ being the C of x for some x in D

Frege exemplifies this method by applying it to the definition of the concept 'direction of a line':[100] on the base of the consideration that, if the line *a* is parallel to the line *b*, then the extension of the concept 'line parallel to line *a*' is equal to the extension of the concept 'line parallel to line *b*', he defines the '*direction of a line a*' as the extension of the concept 'parallel to the line *a*'. He goes on to define the concept of 'number' by analogously substituting concepts for lines and one-to-one correspondence – making objects that fall under one concept and those that fall under the other correspond one-to-one – for parallelism: "the concept F is equinumerous to the concept G whenever there is ... the possibility to put the objects that fall under G and those that fall under F in bijection... Consequently I define: the number that belongs to the concept F is the extension of the concept 'equinumerous to the concept F'. " Next, 'number' is defined as the concept to which an object

[97]"Surely no extensive discussion is necessary to show why I cannot share this view, especially since all the investigations which we have carried out up to this point present nothing but arguments in refutation of it" (*PoA* 124; *PdA* 119).

[98]Cp. Simons 2007, 229 ff.

[99]Frege 1884, §63; Beaney 1997, 110.

[100]Besides 'triangular form.'

belongs iff it is the number of F for some concept F. Finally, by using his (in)famous postulate V, Frege is able to *prove* "Hume's principle":[101]

The number of F's = the number of G's iff F and G are equinumerous. [102]

Husserl's critique of Frege's definition is the same that he already used to criticize the definition of equinumerosity through one-to-one correspondence as well as the theory of equivalence: this method does not define the contents of the concepts 'direction', 'form', 'cardinal number', but only their *extensions*. "We note, however, that all the definitions become correct statements if the concepts to be defined are replaced by their extensions. Correct, but certainly entirely obvious and worthless statements as well."[103]

In an annotation, Husserl adds that "Frege himself seems to have sensed the questionable status of this definition, since he says in a note to it: 'I think that we could simply say 'concept' instead of 'extension of the concept'."[104] However, he does not tell us that in the same note Frege "foresees" the objection that Husserl will move against him with so much vehemence and for so many pages, and Frege claims to be able to refute it, but does not do so because it would lead him "too far." "However, one could object ... that concepts can have the same extension without coinciding. Actually, I am of the opinion that ... this objection can be removed; but that would lead too far here."[105]

In Chapter IX of the *Philosophy of Arithmetic*, "*The Meaning of Numerical Assertions*," Husserl comes back to the Fregean theory of number, this time with the goal of criticizing its characterization of numerical assertions as "predications about a concept."

For Frege, numbers are *objects* while *numerical properties are predicated of concepts*: all statements that express the attribution of a number can be reduced to the form 'the number of Fs is *n*' (e.g., the statement 'Jupiter has four moons' can be reduced to the form 'the number of moons of Jupiter is 4' or – as Frege puts it – "to the concept 'moon of Jupiter' belongs the number four").[106] It follows that, in order to make a numerical assertion (*Anzahlaussage*) it is not necessary to unite certain

[101]Cp. Simons 2007: "As an inspection of Hume shows, it is not close to any principle actually formulated by Hume, and the name follows a somewhat misleading historical footnote to Hume's *Treatise* in *Grundlagen* §63. In fact the principle ... is clearly stated by Cantor in 1895" (246).

[102]Neo-Fregeans like Hale & Wright 2001 try to exorcize the spectre of Russell's paradox by adjoining Hume's principle to second-order logic and showing that the result is a consistent system in which all the fundamental laws of arithmetic are derivable as theorems.

[103]*PoA* 128; *PdA* 122. Cp. Tieszen 1990, 154; Ortiz Hill 1994a, 5 & b, 100.

[104]*PoA* 128, n. 14; *PdA* 122, n. 1. Cp. Ortiz Hill 2002, 96.

[105]Op. cit., 306. Perhaps it is useful to remember that in §2 of *Concept and Object*, while answering a series of objections by Benno Kerry, Frege writes – among other things – that Kerry, falsely, thinks that in the *Grundlagen* the concepts 'concept' and 'extension of a concept' had been identified, and that interpreting them thus would be to misunderstand the content of his book. He then explicitly states: "I merely expressed my view that, in the expression 'the [natural] number that belongs to the concept F is the extension of the concept *equinumerous to the concept F*', the words 'extension of the concept' could be replaced by 'concept'." (Frege 1892, 199; transl. Beaney 187). Cp. Tieszen 1990, 154.

[106]Frege 1884, §56.

objects in a collection, but there has to be a concept that unites them. To the same group of objects can hence belong different numbers, depending on the concept under which they are thought (one of Frege's examples, also reported by Husserl,[107] is: 'here are four companies' / 'here are 500 men').

For Husserl, on the other hand, a numerical assertion always concerns a set of discrete objects.[108] What Frege calls "thinking the same objects under different concepts" for Husserl coincides with the notion of "direction of our interest": the objects alone are not sufficient to constitute a set; it is a change of interest that unifies certain objects rather than others. "With a change of interest is connected a change of concepts under which we distribute the objects into groups and count them."[109] By Frege's lights, subsuming objects under a concept is a precondition of counting, whereas Husserl maintains that the *explicit* subsumption of the objects to be counted under a certain concept does *not* represent a *necessary condition* for counting. For him *numerical properties are predicated of sets.* According to Husserl this is confirmed by linguistic usage: numerical attributes do not have a plural ('four men', 'five trees'), which suggests that it is a collection (*taken as a whole* and not *distributively*) that constitutes the subject of a numerical assertion. While for Frege numerical assertions say something about a concept, for Husserl they rather say something about the extension of a concept:

> On a more exact examination of the issue, it is clear that a number only has a relation to a concept in so far as it counts [the objects of] its extension... Not to the concept ... does the number belong, but rather to its extension.[110]

1.7 Three Further Issues: Unity, Zero and One, Numbers and Numerical Signs

Chapter 8 of the *Philosophy of Arithmetic* ("Discussions Concerning Unity and Multiplicity") and the Appendix ("The Nominalistic Attempts by Helmholtz and Kronecker") address three additional issues that require treatment here:

1. The difference between the concept of 'unity' and the numerical concept 'one'
2. The difference between the nature of *zero* and *one* and that of the other natural numbers
3. The impossibility to interpret numbers as mere signs and the founding and constitutive nature of the conceptual moment for arithmetic

Concerning the first point, Husserl presents arguments for the thesis that the numerical concept 'one', i.e. the meaning of the numeral '1', and the concept of

---

[107] *PoA* 171; *PdA* 163.

[108] "The number is univocally determined when the collection upon which we exercise that abstraction process is determined" (*PoA* 172; *PdA* 163).

[109] *PoA* 172; *PdA* 164.

[110] *PoA* 177; *PdA* 168.

'unit' – as it is to be found in the traditional (Euclidean) definition "a number is a multiplicity of units" – do *not coincide*. The term 'unity' is used to designate things while abstracting from their properties. It has the same logical meaning as the term 'something (*Etwas*)', and this meaning is not to be identified with the numerical concept '*one*'.[111] Numerical concepts are rather, in Cantor's words, "the intellectual image" or "projection" of a certain given set; the product of an act of abstraction which has some kind of existence in our mind.

Husserl maintains that this identification often occurs, and leads to the definition of cardinal numbers (*Anzahlen*) as a sets of ones in the sense of numerical concepts. This error is caused by a more general equivocation which occurs in the case of general and abstract terms, which is due to incorrect *linguistic usage*.[112] According to Husserl, every abstract term is normally used with a double meaning: (i) as name of the abstract concept, (ii) as general name. Given the *Philosophy of Arithmetic*'s specific way of conceiving the relation of common names (general terms) to the objects that constitute their extension (i.e. each term has a meaning – which comes close to a Fregean sense, in virtue of which it applies to the objects of its extension, one at a time),[113] a general term is normally used to designate concrete things and processes. For example, the word 'red' is used to denote the property of being red as well as red objects. An analogous phenomenon occurs with the names 'one' and 'unity'. Husserl's general contention is rather doubtful. As a matter of fact only very few terms in our language do double duty as concrete general terms and as abstract singular terms ('the flag is red' and 'red is my favourite colour'). Pace Husserl, normally the distinction is clearly marked: 'courageous' vs. 'courage', 'wise' vs. 'wisdom', etc.

Let us now proceed to the second of the three points listed at the beginning of this section, i.e. the difference between *zero* and *one* on the one hand and the other numerical concepts on the other. A number for Husserl is the answer to the question 'how many? (*Wieviele?*)', but to this question there can be two kinds of possible answers: *positive* and *negative* answers. The positive answers correspond to all the numbers of the series of naturals except zero and one; the negative answers correspond to zero and one.[114]

---

[111] "Unit (*Einheit*) in *contrast* to multiplicity is not the same as unit *in* the multiplicity. Along with the concept of the multiplicity (or number) the concept of the unit is inseparably given. But in no way this is true of the concept of the number one. The latter is only a later result of technical developments (*ein späteres Kunstprodukt*)" (*PoA* 141; *PdA* 134).

[112] "It will often be necessary for us to lose ourselves in what seem to be linguistic investigations concerning the meanings of the terms in order to put an end to obscurities and misinterpretations of the concepts that interest us. We find ourselves in such a position also with the question that is to employ us now: namely, that about the relationship between the concepts or terms *one* and *unit*" (*PoA* 143; *PdA* 136).

[113] Frege, Letter to Edmund Husserl, 24 May 1891.

[114] "No-many, or no multiplicity [*keine Vielheit*], is not a special case of *many*. *One* object is not a collectivity of objects. Therefore the assertion that there is one thing here is no assertion of number. And likewise, *no* object is not a collectivity, and therefore the assertion that there is no thing here is no assertion of number" (*PoA* 138; *PdA* 131).

Between the positive and negative answers to this question there obtains a difference in *conceptual content*, that logic must not neglect. Linguistically, zero and one behave as numbers, and hence they can be regarded as numerical determinations from the grammatical point of view, but this is not so *from the logical point of view*. The *conceptual* difference between zero and one and all the other numbers should not be ignored: only the latter are numerical concepts in the real and proper sense. "[For the latter] the unity of the concept ... is an *intrinsic* [*innere*] one. They form a logical genus in the narrower sense... [For zero and one] the unity of the numerical concept ... is an *extrinsic* one, established by means of certain relations."[115]

It remains to be explained why zero and one are included in the series of natural numbers. Husserl gives two reasons, both of an essentially formal kind: (i) zero and one are necessary with respect to the *calculistic* aspect of arithmetic, i.e. to the elaboration of the *algorithms* (for instance, think about the role of the zero in the systems of numerical notation of positional kind);[116] (ii) the fundamental numerical relations are *invariant* with respect to the two fields, the *restricted* one (constituted by the numbers from two onward) and the one *expanded* with zero and one. Even if a residue of the different nature of zero and one can be found precisely in the algorithmic-formal laws of the operations, exactly in the "exceptions" that, when zero and one are involved, occur for the general rules valid for arithmetical operations: "The addition of zero does not increase, the subtraction of it does not diminish... Multiplication by one does not proliferate, division by one does not split up, and so on."[117]

Let us now turn to Husserl's third and final point: numbers cannot be considered as mere signs; to interpret them in this sense would mean a misunderstanding of the *symbolical* nature of calculation.

Numerical signs that constitute the series of natural numbers as well as the other number systems have the function of being "surrogates for" or of "representing symbolically" the real and proper numerical concepts, and lack meaning if we disregard their possibility to refer to the latter. The nominalist interpretation for Husserl is a misunderstanding, a quite natural one at that, of the process of symbolization that is at the base of the constitution of numerical concepts as such, and also at the base of the algorithms and the mechanical calculus. The means through which we have arrived at numerical concepts is by representing different sets of concrete objects through certain determinate signs, e.g., bars (11, 111, 1111, ... etc.). In the process of constitution every single object is assigned with a bar and the set of bars resulting from this is considered as the result of the process of abstraction. But the set of bars, the sign, is not the numerical concept itself, it is only a convenient representative of it.

---

[115] *PoA* 141; *PdA* 133.

[116] "Certainly the decimal number system ... would be unthinkable without this momentous expansion of the concept of number" (*PoA* 139; *PdA* 132).

[117] *PoA* 140; *PdA* 133.

"This modelling [*Abbildung*] of each one of the things to be enumerated by means of a uniform sign, meaningless in itself, in fact only mirrors the process that leads from the concrete multiplicity to the number. And *only in so far* as it does so has it sense and significance."[118] The nominalist interpretation misunderstands the sense of this process in the same way as it misunderstands the representative function of the signs; it mixes up the sign with the thing, restricting itself to "enumerating merely as a machine-like exterior process" and hence ending up "totally overlooking the logical content of thought which confers on it justification and value for our mental life."[119]

The point on which Husserl will insist (and to which he will return, with even greater effectiveness and clarity, in Chapter XIII) is that while calculating is an activity that proceeds with signs *and not with concepts*, nevertheless at the end of every calculation the result obtained is the sign *for a numerical concept* and as such it must also be interpreted. On this basis he criticizes not only the nominalist interpretation of numbers commonly attributed to Helmholtz and Kronecker, but also their choice of the concept of *ordinal* number as fundamental concept for arithmetic; thereby justifying 'a posteriori' his own choice of the concept of *Anzahl* as constitutive.[120] Husserl's accusation is that Helmholtz and Kronecker fail to understand the symbolical character of the numerical system and of the techniques of calculation, by trying to find the origin of numerical concepts in the process of computation. For Helmholtz, "every number is determined only by its position in the series of natural numbers"; the series itself is a succession of arbitrary and conventional signs, and all the arithmetical operations are, in their turn, purely operations with signs, while the properties of the operations (commutativity, associativity, ...) are equivalences among sign-complexes. The meaning of each sign is to denote a certain position in the natural ordering of the series from which follows that ordinal numbers are the fundamental numerical concepts. In short, Husserl affirms that Helmholtz confuses cardinal numbers with ordinal numbers, and tries to explain the latter nominalistically as mere signs, while whichever of the two concepts (cardinal or ordinal number) is to be taken as "fundamental concept" for arithmetic, the ordering of the series is determined *by the very nature of the numerical concepts* and does not have anything to do with a conventional sequence of arbitrary signs.[121] Analogous considerations are implied in the refutation of Kronecker's position.

In the second part of the work, Husserl will show that he understands the possibility that the system of symbols can function independently and autonomously from the concepts that it was created to express, but he will always insist

---

[118] *PoA* 135; *PdA* 128.

[119] *PoA* 135; *PdA* 128.

[120] In the Introduction to the *Philosophy of Arithmetic*, Husserl indeed had postponed such a justification to the end of the discussion about the concepts of unity, multiplicity and number, being convinced that the subsequent considerations about the constitution of such concepts would be valid independently of this choice.

[121] Cp. Frege's criticism of Helmoltz in 1903, 139–140.

on requiring, as a *conditio sine qua non*, the *presence of a corresponding system of concepts that, while not being the explicit object of consideration in the calculatorial technique, warrants the meaningfulness of the signs.*

1.8 Arithmetic Does Not Operate with Proper Numerical Concepts

The second part of the *Philosophy of Arithmetic* intends to investigate – from an epistemological point of view – the genesis of the *art of computation* (*computistics*) founded on the concepts of multiplicity, unity, and cardinal number, and moreover to study the relationship of this computistics with institutional arithmetical science.[122]

In Chapter 2 (*Mind and philosophy of number*) of his *Mechanism, Mentalism, and Metamathematics* Judson Webb clearly sets out the conceptual framework by recognizing two different conceptions of arithmetic that he calls respectively 'the *algorithmic* conception of arithmetic' and the '*theoretical* approach to it'.

> [The former] regards the basic operations of arithmetic as algorithms rather then as functions in the modern sense, i.e. as *rules* rather then sets, [it] concentrates on the construction of arithmetic operations and has a practical orientation. [whereas] the more theoretical approach of Frege and Dedekind . . . reduces the notion of number to concepts of pure logic and then concentrates of the proofs of arithmetical propositions. We could express their respective ideals of completeness as 'algorithmic completeness' – which would require an account of all the 'proper mixtures' of arithmetical operations . . . – and 'deductive completeness', which would require an account of all logical axioms. The algorithmic conception tend to stress formalism and concrete symbols while the deductive conception stresses concepts and abstract objects.[123]

As we shall see, the second part of the *Philosophy of Arithmetic* clearly fits into the framework of the 'algorithmic conception of arithmetic'.

Husserl's discussion begins with the observation that the numbers with which professional arithmeticians work are not the numerical concepts, in particular they do not correspond to what the logico-psychological analysis has established numerical concepts to be. These latter, in so far as they are *produced* by a single act of abstraction applied to a concrete set, are 'forms' or 'species' (general concepts) of sets[124] to which additionally is associated a numeral (*Zahlwort*). While it is evident that for the purposes of counting and calculating general 'forms' or 'species' of sets are not used, nevertheless institutional arithmetic seems to proceed with the tacit presupposition that it operates with the real and proper numerical concepts themselves, considering arithmetical operations as operations on concepts. On this view,

---

[122] *PoA* 191; *PdA* 181.

[123] Webb 1980, 44.

[124] I borrow here terminology from Cantor 1991, 365.

addition and subtraction are taken as fundamental operations from which all the others are derivable through specialization. In other words, arithmetical operations are not anything but *particular cases* of the two fundamental activities of addition and subtraction: multiplication a particular case of addition, exponentiation a particular case of multiplication, and so on.

Husserl's argument in the discussion of this "prejudice," which according to him has blocked a a proper philosophical understanding of arithmetic, can be reproduced schematically as follows. If by "operations" are meant *real* activities on the real and proper numerical concepts, the only activities in this sense are addition (*Addition*) and partition or division (*Teilung*), intended respectively as collective connection of the elements of two or more sets (i.e. of the units of two or more numbers) and partition of a set into its subsets. These activities are founded on our representational ability to unite many sets in a single one that comprises them all, and to divide a set into the parts that are its components. However, careful analysis shows that what arithmetic calls "operations" does not correspond at all to the concepts of 'uniting' and 'dividing', or to particular cases of these concepts. For the interpretation described above, for instance, multiplication is a method to make sums of the same *addenda* faster and easier ("to shorten 3 + 3 + 3 + 3, we say four times three, where the addition is tacitly understood").[125] However, if multiplication is nothing but an abbreviated way to write sums, why does arithmetic speak of a *new* operation? An abbreviation in the notation to write sums "may, as such, be very convenient and useful, but it is, after all, no operation."[126] Moreover, taking the "institutional" point of view implies that the concrete executability of the operations depends on the possibility of *effectively executing* the additions and subtractions of the real units that form the foundation of the operations.[127] But arithmetic does not ever actually take into consideration this possibility, nor does it consider as a limit the effective impossibility of operating on the concepts themselves.

The core of the entire argumentation is the following: *We can conceive in a proper and effective way only very small numbers*. Our presentational capacity is limited to such an extent that already for numbers beyond three we cannot distinctly see the real units that effectively compose them.[128] How, then, can this be reconciled with the fact that arithmetical operations also deal with numbers that are much bigger, for which no proper presentation is possible, and furthermore that arithmetic does not consider the incapacity of presenting those numbers as a real problem? In this context, Husserl introduces the notion of *symbolic presentation* and tries to show that the numbers and methods of

---

[125]*PoA* 195–196; *PdA* 185.

[126]*PoA* 196; *PdA* 186.

[127]e.g., $4{\cdot}3 = 3 + 3 + 3 + 3 = 1 + 1 + 1 + 1 + 1 + 1 + 1 + 1 + 1 + 1 + 1 + 1$.

[128]Miller 1982, 9 observes that "... one of the eight theses [Husserl] chose to defend in a formal disputation in the summer of 1887 was the following 'in the authentic sense one can barely count beyond three' (*PoA* 357, *PdA* 339)". Cp. Tieszen 2004, 32.

arithmetic are not properly conceived or real operations on the concepts themselves – as "is assumed inside and outside this science" - but *symbolical numbers and methods*, and that this circumstance determines the sense and scope of all arithmetic.

The conclusion at which he arrives is the following: the genesis of general arithmetic is to be found in the fact that we are almost always forced to limit ourselves to symbolical number presentations. Arithmetic as a whole is nothing but '*a collection of artificial means*' to alleviate the essential incapacity to have a proper presentation, effectively and actively, of all numbers.[129]

> If we had authentic [*eigentliche*] representations of all numbers, as we do of the first ones in the series, then there would be no arithmetic, for it would then be completely superfluous. . . . But in fact we are extremely limited in our representational capacities. That some sort of limits are imposed upon us here lies in the finitude of human nature. Only from an infinite understanding can we expect the authentic representation of *all* numbers; for, surely, therein would ultimately lie the capability of uniting a true infinitude of elements into an explicit representation.[130]

Given that Husserl attributes great importance to symbolic presentations, it is worthwhile to examine their main characteristics, before looking at their particular application to arithmetic and the calculus.

1.9 Symbolic Presentations

Husserl is firmly convinced that *symbolic presenting*, our capacity to refer to things that we do 'intuit directly', plays an essential role on our psychical and intellectual life in general and especially in the constitution of the numerical field: indeed, it is precisely this capacity that constitutes the very possibility of arithmetic. He claims to have taken over the distinction between *proper* (*eigentliche*) and *symbolic* (*symbolische*) presentations from Brentano, partially modifying it.[131] Brentano, in turn, adapted Leibniz's distinction between '*cognitio intuitiva*' and '*cognitio caeca vel symbolica*',[132] and this Leibnizian heritage is also visible

---

[129]Cp. Weyl's remarkable summary of Husserl's *Philosophy of Arithmetic* in his *Habilitationsvortrag* in Göttingen (Weyl 1910, 302).

[130]*PoA* 201–202; *PdA* 193. Cp. Tieszen 1996, 304: "It is a basic epistemological fact, for example, that we are finite beings. Elementary finitary processes such as counting objects in everyday experience, collecting them, or correlating them one-to-one, are clear and familiar to nearly everyone. It is not necessary to know any set theory in order to be able to do these things. And recall Poincaré's characterization of mathematical induction: it is 'only the affirmation of the power of the mind which know it can conceive of the indefinite repetition of the same act, once that act is possible.' On Husserl's view, number theory is founded on processes of this type."

[131]"To [Brentano] I owe the deeper understanding of the vast significance of inauthentic presentations for our whole mental life; this is something which, so far as I can see, no one before him had grasped" (*PoA* 205; *PdA* 193). Cp. Miller 1982, 9.

[132]Leibniz 1684.

in Husserl when he talks about the "relations between blind (i.e. purely symbolical) and intuitive (proper) meaning."[133] The presentation of something is *proper* when it is given to us, "as it were, *in persona*,"[134] e.g., we have a proper presentation of a house when we actually see it. A symbolic presentation is *a presentation through signs* that univocally identify the presented object. In this case the signs are presented properly, while the object is presented indirectly, through the signs. Of the same house we have a symbolic presentation when somebody describes it to us univocally e.g., as "the corner house on such and such side of such and such street."[135]

A symbolic presentation denotes its object in such a way that it can always be identified. We want to be able to speak about the things themselves, about their properties and the relations in which they stand to one another, even though by way of signs. The symbolic presentation has the function of "standing in (*surrogieren*)" for the proper, intuitive one when it is not available. Such a substitution can be temporary, as is the case, e.g., of the house, or permanent, if a proper presentation is forever beyond our cognitive reach; in this latter case the symbolic presentation has the function of a *permanent surrogate* for the proper one, as Husserl puts it. Between the proper presentation of an object and its various improper or symbolic presentations there obtains a relationship of "logical equivalence." On this is founded the possibility of substituting symbolic representations for proper ones in judgements.

Husserl stresses that he has brought forward, in a much stronger way than Brentano, the fact that the symbolic representations *univocally identify* a certain object; and the reason for this clarification is the distinction of symbolic from *general* presentations.[136] In effect, this *univocity* is a determining property in the symbolic construction of the numerical field, which is the problem that we are addressing here.

Symbolic presentations are not just surrogate presentations of direct intuitions: they also have the important function of enabling our instantaneous apprehension of aggregates of objects, even quite numerous ones. Thanks to them, it is possible to isolate groups of objects of the same kind or with a shared characteristic from the totality of the field of our perceptions and to (*symbolically*) form the corresponding aggregate with them.

In short, symbolic presentation is a "presentation through signs"; in general, nothing can be expressed (in a systematical way, for cognitive purposes), and *almost* nothing can be thought, if not by way of signs. This provides for the possibility of arithmetic as a science.

---

133 *LU* II, §14a, 141; *LI* 367.

134 *LU* VI, §45, 144 ; *LI* 785–786.

135 *PoA* 205; *PdA* 194.

136 A general representation cannot represent a particular object; e.g., the representation 'man' cannot be considered an unequivocal sign for a specific individual, e.g., Peter. In order to transform it into such a sign, we have to add to the general representation some distinctive features (*Merkmale*) that unequivocally identify that unique individual.

The analysis of symbolic presentations, insofar as we also find them in the procedures of the calculus used in arithmetic, and, furthermore, the analyses of the processes that lead to the systematic organization of certain specific symbolic presentations in a *numerical system*, are articulated in the following way:

a. An explanation of the possibility for a human intellect to conceive symbolically *greater sets of concrete objects* ("*sensuous sets*").
b. An explanation of the capacity of a human intellect to conceive *infinite* sets and, the decomposition of this conceiving into its components.
c. Constitution of the particular infinite set of the natural numbers on the basis of the results of the preceding point, i.e. using the elementary components of the presentation of an infinite set, and adding to this an "*ordering principle*" for the symbolic number presentations constituted step by step.

1.10 'Sensuous Sets' and Infinite Sets

In Chapter XI, "*Symbolic presentations of multiplicity*" – which is partially an essay in descriptive psychology in the style of Brentano and Stumpf – Husserl deals with the issues mentioned above in (a) and (b). Husserl asks *how we can conceive sets* for which it is not possible to carry out those two fundamental acts that are necessary for the constitution of the concept of set: the 'singular apprehension' (i.e. intuiting each element distinctly for itself) and the 'collective connection' (i.e. having them all present together in one single act). All the sets, called here '*sensuous sets* (*sinnliche Mengen*)', which have so many elements that it is impossible to present them *properly*, are of this kind. In other words, the problem is that of providing a plausible explanation of our capacity to grasp immediately greater collections of sensuous things of the same type (e.g. presentations corresponding to the words 'army', 'crowd, 'flock', etc.). Husserl argues that in the perceptual field there are some elements that are able to exercise a great influence on our interest and attention. These elements, even though constituted by some of the members and relations of the set under consideration, are 'fused together (*verschmolzen*)' and are grasped by consciousness in the same way as a simple quality, though they are not simple at all: Husserl calls these moments 'quasi-qualities of second order'. These 'mediate the association', i.e. they make the instantaneous grasping of sets possible, when it is impossible to see every single element distinctly.

When presented with a sensuous set, e.g. a flight of birds, our consciousness immediately grasps only the quasi-quality, i.e. some elements of the set and the *figural moment* of their distribution, and this *surrogates* for the other elements that cannot be intuited distinctly one by one: in this way we form the unitary presentation 'flight of birds'. That is why we can *symbolically conceive greater sets* composed of things of one kind.

The next step is constituted by the theoretical justification of the possibility, for a human intellect, to present symbolically infinite sets. Symbolic presentations of sensuous sets do not exhaust the extension of the concept 'symbolically constructed set'. Indeed, we speak of sets also when not only the proper presentation of all the elements together, but also the intuition of each of them singly — even taking into account the idealizing cognitive capacity of a human mind — constitute a logical impossibility.[137]

The symbolic expansion of the numerical domain is justified, according to Husserl, by the ability of the human intellect to refer to infinite 'objects': infinite processes, infinite sets, etc.; e.g., the set of points in a line, the set of moments in a temporal interval, and so on. Of course, our presentation of infinite sets is not and cannot be effective: its logical and psychological content is constituted by the presentation of only few elements of the set under consideration, together with the "symbolic presentation of unlimited *process of construction of concepts*." As Tieszen puts it, "this kind of analysis ... brings a constructive turn to the *Philosophy of Arithmetic*. There must be some constraints on the cognitive process of collecting, for example, because we are considering what is possible for *human* collectors, and human collectors are not omniscient."[138] In other words, our general presentation of an infinite set *has an inductive nature* and consists in (i) the presentation of only few elements of the set, (ii) the presentation of a *principle of construction* to obtain all the other elements, and (iii) the certainty that such a process of construction can be carried on indefinitely.

In the concept of infinite set, e.g., that of the natural numbers, we find the proper presentation of "a set in the usual sense – namely the numbers of an initial subsequence of the number sequence ... To this is joined the supplemental representation that this sequence, in view of its principle of formation, can be extended *in infinitum*."[139] With this argumentation, we repeat, Husserl thinks he has theoretically justified the given fact (*Tatsache*) that we have the possibility to conceive infinite sets. All his subsequent explanations about the construction and organization of the numerical system are founded on this.

The reflections about infinite sets in general considered above are elaborated and refined with respect to a specific infinite set, i.e. the natural numbers: we only have proper presentations of very small numbers, and moreover the idea that the restricted field of properly presentable numbers can be extended indefinitely *in a symbolic way*. Hence the problem becomes that of explaining how this idea of a symbolic extension is actually realized. Husserl conceives all numbers, except those intuited directly, as products of a process of *numerical construction* or *constitution*. In the *Philosophy of Arithmetic* there is much talk of *numerical constructions* (*Zahlgebilde*), and the constitution of the numerical field in its totality is considered as a process of conceptual construction that produces the "closed but infinite" totality of the natural numbers. Tieszen's remarks about intuitionistic constructions apply here as well: "the notion of construction here is ambiguous

[137] Cp. Ortiz Hill 1997b 147 & 2002, 82–83 & 84; Tieszen 1994, 332.

[138] Cp. Tieszen 1990, 151.

[139] *PoA* 232; *PdA* 220.

between process and product. In one sense the 'construction' is the cognitive process carried out in time by the subject ...; in the other sense it is the object obtained through this process."[140]

According to Husserl, it is necessary to find an adequate method to execute this construction in an *effective and systematical way*, i.e. a method capable (i) of generating new constructions starting from given ones, of transforming an already constructed concept into a new *distinct* concept; (ii) of warranting *a priori* that every concept generated in this way is *unique*; and (iii) of establishing *a priori* which elements do and do not belong to the generated set.[141]

For the purpose of laying down conditions that the final organization of numerical system must satisfy, several alternative ways of constructing the set of natural numbers – from simpler to more complex – are taken into consideration.

1.11 Unsystematic Number Symbolizations and the Natural Number Series

Faithful to the psychological method chosen as a guide for the theoretical explanation, Husserl begins with the first method of construction that would come to mind if, instead of availing ourselves of the arithmetical science as a well-developed whole, we *really* had only proper numbers available when trying to construct the numerical field in its entirety. Such a method consists in the expression of all numbers through the *additive composition* of proper ones. If properly presentable numbers are those, e.g., from 1 to 10, we would have constructions like '10 + 1', '7 + 5', ..., each conceived as a 'symbolic form' that designates a certain natural number.

The defects of this method of construction are evident: (i) the simple additive composition of proper numbers produces a multiplicity of symbolic forms that is too large, lacking an organizational criterion; (ii) the same number can be designated by different compositions of designations of proper numbers (*no univocity*); and consequently (iii) we have no criterion for immediately recognizing the relationship of each given symbolic number to the others with respect to the order relation ($\leq$) and, furthermore, there is no method to order the generated constructs. The conclusion is that such symbolizations "cannot, with their vague generality, serve the purpose of counting and calculating."

In the light of what has emerged above, some fundamental prerequisites for an adequate symbolic construction of the numerical field become visible. The ultimate organization of the numerical field must be such that (i) we can effectively calculate with all conceivable numbers; (ii) these are univocally classified on the base of the order relation $\leq$; and (iii) the following conditions are met: speed in calculations,

[140] Tieszen 2001, 238.

[141] Ortiz Hill 1997b, 147 & 2002, 83.

ease of distinguishing numerical constructs and univocity in the principle of construction (the same number must not be present more than once, i.e. all must have different names in the construction).

Finally, the chosen solution must match with the computational side of institutional arithmetic. In other words, the definitive organization of the numerical field in a system will constitute the technical side of arithmetic, while the description and explication of the workings of the system will serve as theoretical foundation of the actual arithmetical methods. On Husserl's view, these methods work, but "do not understand themselves," and, when the problems (*Aufgaben*) to be solved are complex enough (as in the case of calculations with the 'imaginary'), this lack of understanding can even cause interruptions in the development of the methodologies of the calculus and, in any case, errors in the correct interpretation of the latter. Therefore Husserl's ultimate aim is to *give a theoretical foundation to the symbolical aspect of arithmetic*, an aspect that functions *autonomously* and that professional mathematicians tend to mistake for arithmetic as such.

A second way of constructing the extension of the properly given numerical field relies on the *natural number series*. Unlike the method of additive composition, the series constitutes a *systematical* extension of the natural numerical field. The principle for the construction of symbolic number forms – the operation of "successor" – is indeed, in this case, strictly univocal and definite at each step. Therefore it is *a priori* certain that to each proper number corresponds only one symbolic form and that the difference in numerical forms implies the difference of the corresponding proper numbers. Finally, because the classifying principle is the order relation, we can simply decide on the base of their position in the series which number is greater.[142]

The methods of symbolically constructing the set of naturals using properly presentable numbers as initial primitive objects and the addition of one unity at each new step as the construction principle, is still not effective: it presupposes a strong idealization of our presentational capacity. In fact, we cannot indefinitely execute the necessary repetitions of the application of the principle and give them a definite order; we lack – as Husserl puts it – the time and spiritual energy required for an activity of this kind, and we also lack the signs to distinguish all the constructs that we produce step by step. It is not enough to have a systematical method to arrive at an extension of the numerical field beyond the one properly given; we also need a method that is effective, able to actually reach numbers that significantly exceed this field. One problem is how to construct an infinite set, another is how to structure it. An adequate method of construction must take into account both these requirements. The defect of the series consists in the fact that it does not have an efficient method to identify the sequentially produced numerical concepts in a manner that it is adequate for the purposes of arithmetical science. The series cannot have a practical and effective use, because we would need a new name for each new numerical construct, and soon the multitude of names would become intractable.

---

[142]Cp. Ortiz Hill 2002, 83.

In principle the natural number series can be continued *in infinitum*, "but for all that, it is actually carried out and given to us only within the limits of what we name (*aber wirklich ausgeführt und gegeben ist sie uns doch nur innerhalb der Grenzen der Benennung*). How are we to hold the uniform steps of number formation distinct in their limitless succession – where each new step indeed presupposes the whole series of earlier ones – without the support of accompanying designations?"[143]

1.12 The Numerical System

The impossibility to solve sufficiently complex calculatory problems with the natural number series shows the need for a more inclusive and powerful conceptual construction, which requires more refined logical tools. Apart from satisfying the prerequisites mentioned above, the expansion of the proper numerical domain must be effected in such a way that all numerical constructs can be obtained through a *few fundamental* signs according to a unitary and easily understandable principle of construction.[144] It is also necessary that the names for the numbers of arithmetic are constructed starting from a few basic names and can be thereby controlled. So Husserl turns to what in current terminology is called 'a system of numeration in a given base'. The essence of such a system consists in the fact that it constructs all numerical concepts using a few elementary concepts and rules for operations.

The base ($1, \ldots, X$) is constituted by the numbers that are properly given to us or by those next to them, that, while not properly presentable, are accessible to us without complex operations or symbolizations.

> "Therefore let us consider the numbers
>
> $1, 2, \ldots, X$
>
> in their natural sequence, as the beginning segment of the system *given to us*; and we then, to start with, attempt novel formations (*Gebilde* – constructs) following the old sequence principle:
>
> $X + 1, X + 1 + 1, X + 1 + 1 + 1, \ldots$"[145]
>
> Instead of these sums, we can "write more simply (*wir zeichnen einfacher*):
>
> $X + 1, X + 2, \ldots, X + X,$
>
> $X + X + 1, X + X + 2, \ldots, X + X + X,$
>
> $\ldots\ldots$"[146]

We can substitute the corresponding numerals taken from the initially fixed segment for all sums of units (having at most X members) in order to have a simpler notation. "But this mode of designation also does not suffice. The further we go the more tedious becomes the designation by the accumulating (*sich anhäufende*) sums of X's. A new means of abbreviation presents itself at this point: the simple

[143] *PoA* 242; *PdA* 229.

[144] Compare Tieszen 2000, 256.

[145] *PoA* 243; *PdA* 230.

[146] *PoA* 244; *PdA* 230–231.

enumeration (*Abzählung*) of the X's leads to the multiplicative symbolization in thought and sign; that is, to: *2X, 3X, 4X* ..."[147]

Using the same kind of argument, Husserl justifies the introduction of the symbolization of the exponential type: the constructs formed by the iterative multiplication of the X (*XX, XXX, XXXX*) proliferate to such an extent that new abbreviations become necessary; counting the factors leads to the operation of exponentiation (X^2, X^3, X^4, ...). We see that the series continues and that the iteration of the last introduced operation always leads to a new operation; however, *for practical reasons and ends* – Husserl maintains – it is *sufficient to stop at the operation of exponentiation*. While the series only has the operation of successor as a method to generate numerical constructs, the system in base X has much more articulated procedures of numerical construction, which *generate and at the same time designate* each number systematically, instead of using a series of repetitions of the number one, starting from the numbers 1, 2, ..., X.

Mathematically, all this means that each number is a "whole, whole-number function (*ganze, ganzzahlige Funktion*)" of a determinate fundamental number (*Grundzahl*) X established conventionally, i.e. that it is symbolically presentable as a *finite* set with the form:

$$\{a_0, a_1X^1, a_2X^2, \ldots, a_nX^n\}$$

with a_i between 0 and X.

> In such a way we have attained to a principle (*Prinzip*) of formation [i.e. construction] for numbers and number signs which actually does satisfy the logical requirements imposed: – It makes possible the systematically uniform continuation, beyond any limit, of the narrow domain of numbers given to us. To accomplish this it requires, through the introduction of the symbolic formation principles of multiplication and exponentiation, no other building blocks than the numbers and signs *1, 2* , ..., X. It encompasses, in concept, the entire domain of number: that is, there is no actual (*wirklich*) number to which there would not correspond, as its symbolic correlate, a wholly determinate systematic formation (*systematische Bildung* – a [symbolic] construct within the system) equivalent to it.[148]

The system and the natural number series are two different ways to construct the same conceptual field (the numerical field) in a systematical way. Both generate all numerical constructs starting from a certain number of basic constructs. The numbers of the series and those of the system are symbolic numerical constructs that have the function to "surrogate" for the real and proper numerical concepts that are not accessible to us. But in doing this the system uses as generating procedures not only the operation of successor but also the arithmetical operations (addition, multiplication, exponentiation, ...). Thereby the system enables us to cope effectively with the numerical field in its totality.

---

[147] *PoA* 244; *PdA* 231.

[148] *PoA* 246; *PdA* 233.

Husserl insists on the fact that the system is not only an instrument, efficient with respect to the economy of signs, for obtaining a system of names for the natural numbers. Because the systematic constructs and those of the series identify the same conceptual field, one can be led to believe that the former is just a more convenient way to designate the numbers of the series. In fact, to every number of the series corresponds a number of the system and, with the latter, is associated its name or designation. For example, with the proper numerical concept *three* is associated a systematical number, 3, which has to be distinguished form its designation, the numeral '3'. The *systematical number is a symbolic concept*, not *a sign*. Ignoring this fact can lead to the conviction that the numbers of the system have the function of being the 'mediators (*Vermittler*)' between the proper number and the corresponding numeral. Since the element of the numerical series cannot be considered as actually given in their totality, to obtain an effective mastery of the numerical field we must consider the systematic constructs as actual concepts whose methods of construction and organization are different from those of the series. As we have already seen, the series is a highly inefficient method to construct the numerical field:

> Only a tiny opening segment of the sequence is given to us. Certainly we can conceptualize the idea (*Idee*) of an unlimited continuation of it, but the *actual* (*wirklich*) continuation, even for only the moderate range involved in the ordinary practice of calculating, already places demands upon our mental capabilities which we cannot fulfil. ... Of course we can form the ideal of an unrestricted continuation of the simple number sequence by correspondingly idealizing our mental capacity. We can, further, think of the sign formations of the number system also as a symbolism for the parallel members of the (ideally expanded) number sequence (*als Signaturen für die parallelen Glieder der (ideell erweiterten) Zahlenreihe*). But one must consider well the fact that these all are only modes for representation and expression, which are inauthentic in the highest degree and have their source (*Quelle*) in the idealizations mentioned. To interpret them in another, more authentic sense would be to distort the entire *sense* of the systematic formation of numbers.[149]

Therefore we have to keep in mind that the system is not a way to designate the concepts of the series in such a way as to solve the problems of practical designation. Rather, it is a *different, alternative* way to construct the same concepts that with the series can be constructed only in principle, and to designate them, this time, through the construction itself.

The distinction between (i) proper numerical concepts, (ii) symbolic concepts of the series, (iii) symbolic concepts belonging to the system and, finally, (iv) their designation, is not easy to grasp because we can refer to all of the above only through the symbolic *notations* (the numerals). However, it is important to stress that numerical constructs, *proper as well as symbolic*, are *concepts*; whereas their *symbolic notations* are *signs*.

To sum up: Husserl characterizes proper concepts as "concepts in themselves (*an sich*)"; they constitute the substrate of the symbolic concepts, but it is not possible to conceive them in a clear and distinct way. The constructs of the series

[149] *PoA* 247–248; *PdA* 233–234.

are symbolic concepts, i.e. presentations that serve as substitutes for the proper inaccessible ones and that make calculations possible. However, the series does not simultaneously provide a system of efficient symbolic notation to designate all its constructs. The constructs of the system are pure symbolic concepts, and their presentation is also highly improper, but this time no difficulties arise because in the signs ('1', ..., 'X'), and in the arithmetical operations we find a univocal way to designate all conceivable constructs. Because of this, the numbers of the system must be considered as "permanent surrogates" of the inaccessible proper numbers.

All numerical constructs outside the system (e.g., those in decimal notation of the form $10 + 5$, $3 \cdot 7$, etc.) form a "problem" that "awaits" a solution, i.e. they must be *reduced* to the corresponding number in the system. Husserl regards systematic numbers as *normal forms* to which all the others must be reduced through the operations.[150] Calculating means: reducing a numerical construct to its normal form, to the number of the system that corresponds to it. For example, in '$49 + 17 = 66$', '$49 + 17$' is one of the possible symbolic forms that serve to represent that given number, and '66', the result, is the normal form, the number of the system to which all other representations of that number are to be reduced. '$49 + 17 = 66$' is also an 'arithmetical proposition (*ein arithmetischer Satz*)' and, appropriately interpreted, becomes an extension of knowledge.

According to Husserl, arithmetical operations are procedures to reduce complex numerical expressions to the corresponding number in normal form, and, *vice versa*, to construct complex expressions starting from numbers given in normal form. One hardly needs to emphasize the depth and "modernity" of this view.

1.13 The Symbolic Aspect of the System

The construction of the numerical field in the form of a system has a characteristic that deserves particular attention. On the one hand, the system produces all (symbolically) conceivable numerical concepts, using only the base (1, 2, ..., X) and the arithmetical operations of addition, multiplication and exponentiation. On the other hand, it produces for all numbers designations using the signs '1', '2', ... 'X' and the signs for the operations of addition, multiplication, and exponentiation.[151]

Husserl observes that if the signs are separated from their conceptual correlates, the symbolic aspect of the system keeps working *autonomously*. The system consists of two correlative structures, a conceptual and a signitive one. The *conceptual* structure is a way to generate new concepts by combining elementary concepts in accordance with certain laws: every number derived from the conceptual

---

[150] Webb 1980 observes that "Husserl's theory of calculation has some of the flavour of Church's calculi of λ-conversion: 'systematic numbers' (e.g. Arabic numerals) result from a series of rule governed 'reductions' of 'unsystematic numbers' (terms compounded out of numerals with function symbols), also called '*symbolische Bildungen*'" (25).

[151] Cp. Hartimo 2007, 288.

structure is a new numerical concept and hence allows for the acquisition of new knowledge. The *signitive* structure is a way of producing signs from signs according to pre-established rules, without any need to refer to their conceptual content.

> Let us abstract from the signification of the designations "1," "2," . . . , "X," as well as from the designations of the operations of addition, multiplication, and exponentiation, and take them as totally arbitrary symbols without signification (as, for example, the counters in a game). Let us replace number definitions and operation rules which are the regular medium of systematic procedure, with corresponding, conventionally fixed formulas expressing the equivalences of sign combinations. One will then recognize that, in this way, there actually originates an independent system of symbols which permits the derivation of sign after sign in a uniform pattern without there ever turning up – nor could there ever, as such, turn up – other sign formations that appear in other circumstances, accompanying a conceptual process, as designations of the concepts here formed.[152]

The system of signs works *mechanically*, it is a 'consequential (*konsequent*)' mechanism that produces symbols automatically. It proceeds in an absolutely independent way with respect to the concepts that it was intended to express. This implies that when counting given sets in practice as well as when constructing numbers through operations, the way of operating that leads to the solution is purely mechanical. The point is that calculating is *not* an *activity with concepts, but with signs.*

Husserl's account of the systematics of signs,[153] and, in particular, the way in which the conception of the system of signs as an autonomously functioning mechanism is presented and detailed testify that he understood very well, and embraced, the results of the process of transformation of algebra that, during the nineteenth century, led to the birth of abstract algebra.[154]

In the work of the English algebraists in Cambridge (C. Babbage, G. Peacock, J. W. Herschel) in the period 1830–1840, the abstract properties of arithmetical operations began to emerge from the numerical substrate as 'autonomous'. The so-called symbolic algebra became an algebraic theory of magnitudes in general: on the one hand, it assumed as principles (and hence as rules of calculus) the laws that apply to the usual arithmetical operations, while, on the other hand, it eliminated the restriction concerning their exclusive applicability to natural numbers.

Already here we have a distinction between (i) a symbolic aspect, i.e. a system of formal laws of connection and of abstract algorithms of computation, by means of which conclusions are drawn in a deductive-algorithmic way,[155] and (ii) the possible systems of entities that can satisfy such formal conditions. With the same symbolic system we can provide a unitary treatment for systems of heterogeneous entities that manifest a similar structural behaviour.

With the contributions of scholars such as W. R. Hamilton, H. Grassmann, and A. Cayley there was a progressive distancing from the idea of algebra as "symbolic

---

[152] *PoA* 251–252; *PdA* 237–238.

[153] *PoA* 251 ff.; *PdA* 237 ff.

[154] Casari 2000, 105.

[155] Casari 1973, 8–9.

algebra of magnitudes," which will culminate in the explicit disengagement of algebraic research from the quantitative dimension with George Boole and his creation of an algebra of logic. Algebra no longer only treats numbers or magnitudes, but also propositions, concepts, and, in general, *qualitative data*. The laws under which they fall are independent from any specific interpretation of the symbolism, and the structural properties of the operations that are reflected in such laws are unleashed from numerical elements and assume the character of abstract algorithmic procedures for "calculations" performed with symbolic expressions.[156] The explicit separation between laws of calculus – purely *formal* laws – and their interpretations is more or less the distinctive trait of modern abstract algebra and mathematics. In Boole's words:

> The validity of the process of analysis does not depend upon the interpretation of the symbols which are employed, but solely upon the laws of their combination. Every system of interpretation which does not affect the truth of the relations supposed, is equally admissible, and it is thus that the same process may, under one scheme of interpretation, represent the solution of a question on the properties of numbers, under another, that of a geometrical problem, and under a third, that of a problem of dynamics or optics.[157]

A similar statement about the importance of the *polyvalence* of *symbolic-calculistic* systems can be found in the *Philosophy of Arithmetic*.[158]

In spite of this development of the calculus, it has to be kept in mind that in the case of the numerical system the laws that regulate the symbols actually describe a well-determined reality, that of proper numerical concepts and of the laws regarding the conceptual operations on these concepts; therefore, at least in the case of numerical concepts, the objective referent is not eliminated. Husserl is always entirely aware of the conceptual basis of the numerical mechanism, of the thought that constitutes and accompanies it, and maintains that there must be a conceptual content that guarantees the meaningfulness of mechanical operations with signs, even if *we can manipulate the signs* without attending to the concepts that are their meanings.[159]

In the *Philosophy of Arithmetic* the system of signs has its foundation in conceptual operations. To every conceptual operation corresponds an operation with signs. A relation subsists between concept and sign that Husserl calls *equivalence*, and the system of signs cannot be established without explicating this relationship.[160] The coherence of the conceptual operations is what warrants the correct functioning of the signitive structure. However, in the first case, the transformations happen on the basis of conceptual *knowledge*, while in the second case the transformations of the signs proceed on their own, indeed according to certain types (*Typen*), but extrinsical types, according to a fixed template (*schablonenhaft*).

---

[156]See Cantini 1979, 41 ff.

[157]Boole 1847. Cp. also Webb 1980, 79; Hartimo, 2007, 285 ff.

[158]*PoA* 273; PdA 258.

[159]See e.g. Tieszen 1996, 312–313 & 2000, 9.

[160]Cp. Centrone 2005.

In other words, what happens on the level of signs is legitimated by what happens on the level of proper concepts, and for Husserl this depends on the fact that the signs are construed as expressing concepts and that the system of concepts, as well as the operations on them, are consistent (*konsequent*). It is the exactness of the conceptual operations that determines the exactness of the system of signs, even if the latter can then be considered completely independently from its substratum, the concepts.

1.14 The Concept of Computation

In the last chapter of the *Philosophy of Arithmetic* ("*The Logical Sources of Arithmetic*"), Husserl distinguishes *arithmetic* from *the art of computation* (*Rechenkunst*, computistic). The latter is conceived as the technical side of the methods of the former, on the basis of the distinction between concepts and signs and, moreover, on the basis of the peculiar property of the autonomous functioning of the system of signs.

Arithmetic is characterized as the science of *numbers*, or, more precisely, as the science of *numerical relations*, as it does not have as object the specific properties of single numbers, but numbers in so far as they are identified through certain relations or complexes of relations with other numbers. The task of suitably characterizing the art of computation is more delicate and complex. It is necessary, first, to consider the various meanings of the concept of *computation* (*rechnen*) in order to arrive at a determination of this concept that enables us to distinguish arithmetic from the art of computation.

Husserl distinguishes between various meanings of 'calculation', one broader and two narrower ones. In the *first*, broader sense, calculating is *any kind of derivation of numbers from numbers*. Within this characterization, the *method of derivation* is not qualified, since it can be *conceptual-abstract* as well as *signitive-concrete*. If it is conceptual, new numerical concepts are derived from proper or symbolic numerical concepts on the basis of certain real or symbolic operations *that are conceptual as well*. The signs, in this sense of calculation, have only a subsidiary role. This implies that if we take 'calculation' in this first sense, it is not possible to exclude from the notion of 'computation' the addition and partition of aggregates, and, consequently, there is no real distinction between arithmetic and the art of computation: arithmetic would be the science of numerical relations, and "computistics" would be the art of calculating, i.e. of deriving numbers from numbers according to certain known relations *through conceptual operations*.

In a *second*, more restricted sense of calculation, the method of derivation is required to be sensuous, i.e. it operates on signs according to pre-established rules. Under this reading, "calculating" means "deriving numbers from numbers through operations on sensuous signs". In comparison with the restricted sense of calculation, the method of conceptual derivation is *less general*, and, moreover, makes calculation a long and complex operation. By contrast, the method of sensuous signs, besides being more practical and more functional, is also more general and *all-inclusive*, i.e. *universally applicable* and potentially able to solve any conceivable problem. "The method of sensuous signs is, therefore, *the* logical method for arithmetic." Ultimately, it makes the method of conceptual derivation superfluous. However, Husserl maintains that for a general acknowledgment of this view we still lack a logic of symbolic methods of knowledge.[161]

Even if we take 'calculating' in the last interpretation, in the sense of "deriving numbers from numbers with a signitive method", we would not yet be able to make a conceptual distinction between arithmetic and computistics because, under this reading, too, arithmetic has as its object the procedures of the calculus and cannot be distinguished from a computational methodology.

Calculating can also be interpreted in a *third* sense, different from these other two, i.e. as deriving signs from signs following formal rules. From the conception of calculating – this is Husserl's thesis – must be excluded not only the operation with proper numerical concepts, but also that with symbolic numerical concepts:[162] calculating must be *exclusively an activity with signs*. This new sense of calculating is obtained *via* the consideration that in the symbolic moment of the numerical system we can abstract from the possible structures that it can be applied to. In Husserl's words: "One can ... conceive of calculation as *any rule-governed mode of derivation of signs from signs within any algorithmic sign-system according to the "laws" – or better: the conventions – for combination, separation, and transformation peculiar to that system.*"[163]

With this third sense of 'calculation' we have obtained a true and proper characterization of the *formal-algorithmic* method.[164] An algorithm is in fact a mechanical procedure that operates on configurations of (sensuous) signs according to certain formal rules. Calculating, now, means deriving signs *from signs* according to pre-established formal rules. Husserl attributes great importance to this third concept of 'calculation', since it makes possible an exact separation of the various *"logical" moments* that are involved in every derivation of numbers from numbers. "Each solution obviously decomposes into one calculational part and two conceptual parts: *conversion of the initial thoughts into signs – calculation – and conversion of the resulting signs back into thoughts.*"[165]

---

[161]"Most researchers – guided by the general prejudice that every scientific methodology operates with the respective intended concepts – have also held the arithmetical operations to be abstract-conceptual, in spite of all clear indications" (*PoA* 272; *PdA* 257).

[162]Indeed, in both earlier definitions, calculating was defined as a deriving of numbers from *numbers*, and by numbers were meant the *numerical concepts*.

[163]*PoA* 273; *PdA* 258 (italics in the original). Cp. Hartimo, 2007, 289 f.

[164]As Webb 1980 puts it: "Husserl ... attempted a complete development of the algorithmic conception of arithmetic, which required "*die logische Untersuchung des arithmetischen Algorithmus*". The notion of algorithm, Husserl felt, had to be bound up with that of a 'mechanical process'" (24–25).

[165]*PoA* 273; *PdA* 258 (italics in the original). Husserl does not fail to stress how important a good *choice of the system of signs* is, in terms of efficiency, for all three of these phases (encoding – calculation – decoding) of the solution of a problem.

With respect to arithmetic the conversion of numerical concepts into signs consists in abstracting from the concepts (because sign and concept cannot be properly separated in arithmetic). The importance of the concept of computation on signs is determined also by the fact that it sensitive to the developments in logical-abstract algebra. The system of signs is indeed *polyvalent*: it allows a uniform treatment of "similar" situations or, equivalently, a single computational system permits the mastery of more than one conceptual system. "It is a fact highly significant for the deeper understanding of mathematics that one and the same system of symbols can serve in *several* conceptual systems which, different as to their content, exhibit analogies solely in their structural form. They are then, as we say, governed by the same calculational system."[166]

Taking 'calculation' in the third sense we can finally distinguish arithmetic as deductive science from computational science. Arithmetic is the art of arithmetical knowledge while computational science is its technical side.[167] However, the method of computation of "signitive" kind must be founded on the conceptual moment, i.e. (i) on proper or symbolic numerical concepts that are given to us in the forms of the natural series or of a numerical system, and (ii) on their *forms of composition* that enable us to obtain new numbers from given numbers and that, within a numerical system, coincide with the arithmetical operations.

1.15 The Fundamental Task of Arithmetic

As we saw, in a numerical system there can be different symbolic configurations that designate the same number. Every "non-systematic" complex of symbols must be considered as the presentation of a specific "arithmetical problem," whose solution consists precisely in the *reduction* of that complex numerical "term" (as we would say today) to its *normal or canonical form*. Each non-canonically designated number is "equal" to a systematic number in the sense that they both correspond to the same proper numerical concept.[168]

Systematic numbers must be considered as *ultimate* concepts, because they are indispensable surrogates for the proper concepts that are inaccessible to us. All acceptable numerical forms (terms) are either *canonical* forms or reducible to such forms. Their composition by way of the operations allows us to construct the numerical field in its totality.

---

[166] *PoA* 273; *PdA* 258 (italics in the original).

[167] "If we loose the number signs from their conceptual correlates, and work out, totally unconcerned with conceptual application, the technical methods which the sign system permits, then we have extracted the pure calculational mechanism that underlies arithmetic and constitutes the technical aspect of its methodology" (*PoA* 274; *PdA* 259).

[168] "To each non-systematic number there corresponds a univocally determinate systematic number that is equal to it, i.e. one which symbolizes the same authentic [i.e. proper] number concept" (*PoA* 276; *PdA* 261). Cp. Hartimo 2007, 290.

The idea that underlies this reduction is the following: with the system we have a general and exact numerical classification such that the numbers of the system can be directly and immediately compared according to the relation of order ($\leq$). Presented with two complex numerical forms we cannot immediately decide which one is greater and which smaller. So the method to achieve this is to reduce them both to their corresponding canonical forms and then compare these.

Keeping these considerations in mind, let us try to understand the sense and scope of Husserl's proposal to establish "the fundamental task of arithmetic," i.e. to explicate the proper function of arithmetic as a science once it has been distinguished from calculational methodology. In Husserl's words, the first basic task of arithmetic, articulated in two sub-problems is:

(i) "to separate all conceivable symbolic modes of formation of numbers into their distinct types,"[169] where the 'symbolic modes of formation (or construction) of numbers' are the arithmetical operations, while the term 'type' indicates the kind of composition (additive, multiplicative) that is used.

(ii) "to discover for each type the methods that are reliable and as simple as possible for carrying out that reduction." That is, *for each conceivable operation arithmetic must find a method of calculation, an efficient algorithmic procedure, to execute it.*[170] In short, Husserl has divided the numerical concepts into proper and symbolic and has shown that the numbers of arithmetic are symbolic concepts organized in a system that operates according to certain rules. He has then distinguished the signs from the concepts, showing that the signs constitute a "mechanism" that works autonomously and correctly, without conceptual reference. The system of signs constitutes the "technical" aspect of arithmetic, and in order to function it does not need a further conceptual clarification concerning its signs: it is the set of computational procedures that the professional mathematician (arithmetician) uses. Husserl then returns to the consideration of symbolic concepts of the system, characterizing them as 'numbers in normal form', as 'fixed samples (*feste Etalons*)', to which all other conceivable numerical constructs must be reducible. The constitution of the numerical field is obtained by composition: starting from certain basic elements all the others are constructed following certain pre-established rules. The procedures to construct the numerical field in its totality are the arithmetical operations. Finally, the basic task of arithmetic is characterized as that of finding ever-new arithmetical operations to reduce any unsystematic symbolical sign-configurations to their corresponding systematic one, and to find ever more efficient procedures of calculation to execute these operations.

Husserl explicitly states that the analyses of the last part of the *Philosophy of Arithmetic* aim at sketching the idea of a universal arithmetic (notwithstanding the fact that all considerations refer to the theory of finite cardinal numbers). As we will

---

[169] *PoA* 277; *PdA* 262.

[170] Cp. Webb 1980 25.

try to show in the next section, these analyses actually constitute a first attempt to circumscribe in a rigorous mathematical way a class of arithmetical operations that Husserl calls 'totality of all conceivable arithmetical operations'. More precisely, our thesis is that he has a clear intuition of that class of functions, which, in current logical terminology, is known as 'class of partially computable numerical functions'.[171] The problem of a rigorous characterization of this class – it is useful to remember – was explicitly and systematically tackled only in the 1930s, in the context of the theory of effective computability in mathematics (in the works of A.M. Turing, A. Church, K. Gödel, S.C. Kleene, and others).

1.16 The Taxonomy of Arithmetical Operations

The last four paragraphs of the *Philosophy of Arithmetic* are centred on a detailed taxonomy of arithmetical operations. Husserl first discusses the four *elementary* arithmetical operations (*elementare arithmetische Operationen*) – *addition, multiplication, subtraction* and *division* – and then he takes into account a number of *generative* procedures which produce new operations (*höhere Operationen*) from given ones. As to the former, Husserl's main concern is, on the one hand, to stress again the symbolic nature of these operations as methods to reduce non-systematic constructs to normal forms, and consequently, on the other hand, to give evidence of their *computability*.

Let us consider in detail *addition*. If addition were an operation performed with or on proper numerical concepts (sets of units) adding two numbers would mean uniting two or more sets of units in a new set. It would not be possible to speak of a *rule* to perform the addition. By contrast, taking addition to be a certain method of reduction of non-systematic constructs to their normal form, we can very well speak of a rule to perform this reduction. If we take the series of naturals as our numerical system, adding b to a, $(a + b)$, means adding b units to a – and this is the best method of calculation that we have in this case. In the system in base X, instead, there are more efficient procedures to calculate a sum, e.g., summing by columns. This procedure allows reducing any addition to a series of elementary additions; moreover, Husserl adds, it demonstrates how the requested number is constructed by the numbers that are its parts: "through these same partial numbers ... those elemental additions ... are univocally determined."[172]

Similar considerations are made for the remaining three elementary operations, i.e. multiplication, subtraction, and division. Husserl presents specific examples of computing algorithms and even discusses aspects such as complexity and time spent computing these algorithms (e.g., the algorithm to execute division is more complex than the one which computes the other elementary arithmetical

[171]For a precise definition of the notion of *partial* numerical function and for a formal reconstruction of Husserl's attempt see Appendix 1 below.

[172]*PoA* 282; *PdA* 267.

operations).[173] Also, he dwells on the fact that subtraction and division are the first and easiest examples of numerical operations that are *computable*, but not *total* (i.e. not defined with respect to the whole domain, as addition and multiplication are); they are *partial* operations, in Husserl's words, "obviously the problem does not always have a sense and solution."[174]

A special point is raised concerning multiplication. Husserl observes that only by appealing to the concept of 'symbolic-algorithmic system' one can understand the *proper nature* of this operation. Under the *conceptual* interpretation of numbers and arithmetical operations, multiplication "is a problem," because it appears to be a special case of addition, in which the addenda are equal, while arithmetic takes it to be a "new operation." Now, within a formal-algorithmic system multiplication *is* a new operation. In fact,

> a major abbreviation is already brought about through the multiplicative mode of representation and designation, in that the *number of the summands is introduced (herangezogen) as a means of symbolization*... The problem which multiplication solves consists in this: to calculate the product ... solely from the multiplicand ... and the multiplier ... *without having to actually carry out the addition*, or even to begin it.[175]

We think it is important here to stress the jump, both conceptually and in terms of arithmetical complexity, *from* the *determinate* iterations of addition (the unary operations $x \cdot 2 = x + x =$ two-times x; $x \cdot 3 = x + x + x =$ 3-times x; in general, $x \cdot n = x + x + \cdots + x$ n-times, where the multiplier n is a natural number which is *determinate*), *to* the *binary* operation of multiplication, $x \cdot y$, in which the multiplier y is a variable. The first (used by Husserl in the constitution of the numerical system in base X) are *explicitly* defined in terms of addition, whereas multiplication is not and cannot be so defined. To get a "measure" of this gap in complexity, contrast the *decidability* of the theory of addition of natural numbers (that is, the theory of the structure $\langle N, +, 0, 1\rangle$)[176] with the *undecidability* of the theory of addition *and* multiplication (that is, the theory of the structure $\langle N, +, \cdot, 0, 1\rangle$).[177]

Husserl's full awareness of the matter is also confirmed in a 1891 manuscript with the title "*On the Concept of the Operation*".[178] Here he succeeds in highlighting with extreme clarity that "abstracting from the nature of the domain" is a

---

[173]The method of finding the systematic number corresponding to the construction $a{:}b$ consists in reducing every division to a series of elementary divisions. Nevertheless, while for calculating addition and multiplication it is sufficient to use a table of all elementary additions and multiplications between an i and a j belonging to $\{0, 1, \ldots, X\}$, for calculating division we need a table of all elementary divisions of the form $a{:}c$ where c is a number between 1 and X and a is a "two-digit" number with respect to the basis X.

[174]*PoA* 285; *PdA* 269. As Webb 1980 rightly says "especially remarkable is [Husserl's] suggestion that the question whether an arbitrary '*Rechnungsaufgabe*' is always defined (*Bedeutung haben*) for any number will require a deep analysis" (25).

[175]*PoA* 283; *PdA* 268.

[176]Presburger 1929.

[177]Church 1936.

[178]*PoA* 385–408; *PdA* 408–429.

necessary condition for operating *formally*. The argument is the following: we could, at first sight, think that multiplication is *reducible* to addition, in other words, that it can be formally defined from it. Now, even though it is correct to say that we can *formally* express $a \cdot 1, a \cdot 2, a \cdot 3, \ldots$, additively, as $a, a + a, a + a + a, \ldots$, *we cannot* legitimately consider

$$a \cdot b =_{\text{df}} a + a + a + \ldots b\text{-times}$$

as a *formal definition* of $a \cdot b$. This tentative definition, in fact, has no sense *if we do not already know* that b is a *natural number*. The point is that in order to be *formal*, a definition and, more generally, a proof *must abstract from the concrete domain* (in this case from that of natural numbers). The same applies to the numerical operation of *raising to a power*: in this case, too

$$a^b =_{\text{df}} a \cdot a \cdot a \cdot \ldots b\text{-times}$$

is not a *formal* definition of this operation.

> Why, then, consider multiplication as a new operation? The distinction lies in the fact that, in order to prove the laws of multiplication, we must go back to the natural number or cardinal as a sum or equality (*Gleichheit*) of units, whereas this is not necessary with the proof of the generalized law of association. *All deductions that do not go back to the concept of the domain are formal*... A determination (*Bestimmung*) is a *formal consequence of certain presupposed ones* if it can be formed from them *without ever having to have recourse to the nature of the domain.* In this sense $a + a$, $(a + a) + a$, etc., are formal, but not $a + a + \ldots b$-times. *For this determination loses its sense if I do not think of the fact that* b *is a number*. The same holds for raising to a power... And because this is so, the propositions about the new operations also cannot be discovered from those about the old ones without recourse to the number concepts.[179]

Let us go back to the presentation of the taxonomy of operations as set out by Husserl. Having fixed the four elementary operations as a basis, he proceeds to isolate – in the three final paragraphs: "*The Higher Operations*," "*Mixing of Operations*," "*The Indirect Characterization of Numbers by Means of Equations*" – a number of different methods by which new operations can be generated. In the light of the "fundamental task" of arithmetic the aim is that these generation procedures should prove sufficient to generate *all conceivable forms to determine new numbers*.[180]

A first generation procedure arises from the observation of mutual relations between operations:

> ... a sum of equal addends (thus the cumulative iteration of one and the same number) has yielded, through the counting up of the occurrences of its repeated term, a new means of symbolic number formation: b times a. We have obtained the *product* representation. But a product of equal factors (thus the multiplicative iteration of one and the same number)

[179] *PoA* 407–408 (my emphasis); *PdA* 429.

[180] Cp. Webb 1980, 25.

> then provides, once again, through the counting up of the occurrences of the repeated factor, a means of abbreviated and indirect number characterization. We obtain the *power* concept, a^b. And one easily sees that this new type of symbolic number formation has a sense for any pair of numbers a and b, i.e. it characterizes a wholly determinate number. In the same way we can continue on: through counting how often a number has been iteratively raised to a power there arises a new type of symbolic number characterization, that of *elevation*; through the counting up of iterated elevations, again a new one; and so on *in infinitum*.[181]

Hence, starting from addition, the *iteration* of the last operation introduced leads to a new operation, and this procedure can be iterated an arbitrary number of times.[182] In other words, Husserl has a clear vision of the *succession* of operations[183] later known as 'Ackermann's succession', that is, the infinite succession

$$f_0, f_1, f_2, f_3, \ldots$$

of binary, total numerical functions defined as follows: f_0 is *addition* (which can be defined by primitive recursion[184] from the successor operation), f_1 is *multiplication* (which in turn can be defined by primitive recursion from the addition operation) and, for $k \geq 1, f_{k+1}$ is the function which is defined as follows by primitive recursion from f_k:

$$f_{k+1}(x,0) = 1$$
$$f_{k+1}(x,y+1) = f_k(x, f_{k+1}(x,y))$$

In this way, we get an infinite succession of binary functions, each of which, with the exception of the initial function f_0, is defined in terms of the immediately preceding function by means of the primitive recursion schema.[185] We may thus

181 *PoA* 292; *PdA* 276–277.

182 Webb 1980 points out that H. Grassman "was the first mathematician both to approach arithmetic axiomatically and to employ recursive definition for the basic arithmetic operations. ... Recursive definitions for the basic arithmetic operations began to appear frequently in the literature after Grassman..." (44).

183 For this interpretation see also Casari 1991.

184 See Appendix 1 for a precise definition. Intuitively, a primitive recursive definition of a unary function f (the n-ary case, with $n > 1$, being analogous) consists of (i) an explicit definition of the value of f for the argument 0, and (ii) the definition of the value of f for an arbitrary argument distinct from 0, i.e. for an argument of the form $x + 1$, in terms of the value that f assumes for the argument x.

185 It is reasonable to maintain that Husserl's theorizations here do not go beyond 'Ackermann's succession'. There is no evidence at all that Husserl had realized the gap, both from the conceptual point of view and in terms of arithmetical complexity, between the succession $f_0, f_1, f_2, \ldots$ and the so-called 'Ackermann's function', i.e. the ternary function A defined by:
$A(n,m,r) = f_n(m,r)$
A is clearly a computable function, due to the fact that each function in Ackermann's succession is computable; one can prove however (as Ackermann did) that A is not primitive recursive. A

conclude that the *primitive recursion schema*, even if it is not explicitly theorized, is clearly exemplified by the above *succession of operations.*[186]

A second way to get new operations is to find and identify the *inverse* of each given operation. Each operation belonging to the infinite succession considered above produces only one inverse if it is commutative, and two inverse operations if it is not; e.g., from exponentiation the two operations of *root extraction* and *logarithm* are generated, whereas from multiplication, which is commutative, only division is generated.

> As the concept of product led to the inverse concept of the quotient, so also each of these new forms leads to corresponding inverses. If, for example, the power concept is established, then the symbolic formation a^b points to a certain number c, where $a^b = c$. But now, in virtue of precisely the same relation, b also in a certain way is characterized by a and c, and likewise a by b and c. b is characterized as the number of multiplicative iterations of a which is equivalent to the number c; and a as the number which multiplicatively iterated b times yields the number c. *We therefore have here acquired two new ways of indirectly symbolizing number formations (in symbols, $^b\log a$ and $^b\sqrt{c}$), through the inversion (Umkehrung) of the relationship defining the concept of power. And in the same manner each further member*[187] *in the above sequence of number characterizations obviously supplies, through inversion of its definition, a new pair of characterizations.*[188]

As we see, Husserl considers here the step from a total *not necessarily unary* function *f* (as already stated, all functions of the above succession are binary and total) to its inverse (or inverses); and he expressly poses the problem of the partiality or non-partiality of functions obtained by inversion ("whether ... the problems here characterized have a signification under all circumstances, i.e. for any arbitrary pairs of numbers a, b ..."[189]): the fact of not imposing as a condition the *surjectivity* of the function to invert implies that the functions obtained by inversion can be *partial*, which is the case for the examples mentioned by Husserl: *division*, the *logarithm* and the *root.*

As in the case of elementary operations, also for the *higher* operations considered up to now – i.e. those belonging to Ackermann's succession and those obtained by inversion – Husserl discusses the problem of how to calculate them, in the sense

recursive definition of *A* requires essentially a nested double induction, which does not fit into the primitive recursion schema.

186 Webb 1980 recalls that also Dedekind used "recursion to provide a precise mathematical basis for the systematic introduction of new arithmetical operations". Furthermore, concerning Ackermann's sequence Webb rightly stresses: "Of course, none of these f_k, beyond f_2 [notice that in Webb's definition f_0 is taken to be the successor function] have any use in the market place, as already by f_4, called 'elevation' by Husserl, we encounter a growth rate so steep as to make its calculation a practical impossibility for all but the smallest arguments. By f_5 even our standard notation conventions begin to buckle, while for f_6 we presumably will have to remain for ever content with a recursion formula [of the kind exemplified by the scheme] as our only feasible description of it" (51–52).

187 Read: each operation.

188 *PoA* 293 (my emphasis); *PdA* 277.

189 *PoA* 293; *PdA* 277–278.

of identifying one or more procedures as simple and *efficient* as possible for the reduction to the normal form of the 'terms' containing them. He suggests, in this case, a combination of the method already described for elementary operations (exemplified by the reduction of the sum of two big numbers to 'elementary' sums, i.e. sums of numbers between 1 and *X*) with one that consists in the reduction to operations of a lower level. The discussion of another general issue, i.e. the assessment whether or not certain important properties are valid for the operations that are constructed step by step, e.g., *commutativity* ("whether or not they are affected by interchanging the numerical values [*Zahlenwerte*] without changing the form of the combination"[190]), is connected with this problem: it is apparent that the knowledge of *structural* properties that govern the behaviour of the operations, in this case commutativity, is clearly relevant with respect to possible gains in the efficiency of the calculations.

Among the 'conceivable arithmetical operations' we find also 'compositions of operations (*Operationsmischungen*)'. These are operations formed by the composition of other operations that have a lower degree of complexity and behave as primitive elements:

> But with the number compositions and the corresponding operations taken into consideration up to now, the totality of those that are in general conceivable is still not exhausted. *There is added the entire manifold of new forms that arise from combination of the ones already formed by using them as their basic elements. There arise problems such as, for example, multiplying a sum by a number, dividing a product by a sum, raising a quotient to a power, etc.*[191]

The idea here is that of "putting together many functions," i.e. to move, for example, from a function *f* and a function *g* (which for simplicity we will consider to be unary) to the function *f*(*g*(x)). Each 'composition' constitutes a *method* for generating new numerical constructions starting from given ones and, *vice versa*, to reduce complicated numerical constructions to their *normal form*.

With regard to the general problem we are discussing here, *identifying all conceivable forms to determine new numbers*, Husserl points out that there is a final case that should be considered. In those discussed up to now, we have found systematic numbers symbolically determined by 'complexes' made up of some given numbers and of some operations, and arithmetic has the task to reduce every similar 'complex' that is different from the canonical one to its canonical form, and this has to be done in the most efficient way in terms of speed of calculation. Nevertheless, a number can also be determined *implicitly*, by an *equation* or a *system of equations*, of which it is the only solution. "...numbers can also be defined by *equations* .... Finally, there is yet to be mentioned the possibility that a number is defined by a *system of equations*, rather than by a single equation."[192]

In the first case, we speak of *direct operations* or the *direct determination of a number:* the value is calculated directly from the data. The second case is that of

[190] *PoA* 293; *PdA* 277–278.

[191] *PoA* 294 (my emphasis); *PdA* 278.

[192] *PoA* 297; *PdA* 281.

indirect numerical characterization, as Husserl puts it, i.e. when a number is symbolically defined as "an unknown constituent of … a precisely characterized structure of numbers." The value of the unknown component is calculated by "unravelling (*aufwickeln*)" relations that occur between the unknown and the known numbers: "here we now have before us a far more difficult problem: namely, that of *unravelling complicated number interrelationships into which the unknown number itself is interwoven.*"[193] In general, an equation in one unknown x can be represented in the form $t = c$, where c is a known number and t is a combination of operations made up of known numbers a, b, … and the unknown x; in the case in which there is a unique solution n to the equation (i.e. a unique number n such that for $x = n$ the equation becomes true), this n is in fact a function of the parameters $a, b, c, \ldots$.

Of particular interest is the thesis that *equations* and *systems of equations*[194] are *a generalization of that particular kind of indirect numerical characterization that is obtained by inverse operations*. The specific examples of inversion considered by Husserl (i.e. the inversion of the direct operations of addition, multiplication and power which gives rise, respectively, to subtraction, division, root extraction and logarithm[195]) can also be expressed as equations; more exactly, each of these operations is a specific case of the solution of an equation. *Subtraction* solves (if this is possible, by remaining in the domain of natural numbers) the equation $a + x = b$ (for a and b given numbers, and unknown x); the same holds for *division*, which solves the equation $a = b \cdot x$, and for the two inverses of the power: "*b-th root of a = the unique number x, if it exists, that solves the equation* $x^b = a$" and "*base-a logarithm of x = the unique number x, if it exists, that solves the equation* $a^x = b$". In Husserl's words: "one immediately sees that in all of these cases we have a generalization of that type of indirect characterization of number which we have observed in every 'inverse' number formation. In the first sequence of number formations a number x was defined by means of the combinations:

$$a + b, a \cdot b, a^b, \text{etc.};$$

and in the second sequence by means of the conditions:

$$a + x = b; \quad a \cdot x = b; \quad a^x = b; \quad x^a = b; \text{etc.}$$ [196]

---

[193] *PoA* 297; *PdA* 281.

[194] As regards *systems of equations*, we observe that it is always possible to associate with every system of equations an appropriate equation, equivalent to the first in the sense that a number x solves each equation of the system if and only if it solves the equation associated with the system. The case in which a number is determined by a system of equations, rather than by one, is therefore, as Husserl stresses, a case that "in spite of the greater degree of complication, offers nothing essentially new from the logical perspective" (*PoA* 297; *PdA* 281).

[195] But we can presume that Husserl also has in mind inversion, for example, of the direct operation of *tetration* (or *super-exponentiation*), i.e. the one that immediately follows the *power* function in Ackermann's succession.

[196] *PoA* 297; *PdA* 281–282.

Moreover, Husserl observes that the inversion is not limited to operations which are "not-composed" (as they are with *Ackermann's succession*): "but now through combinatorial linkage of these[197] there are also to be constructed other complicated conditions of the same character: e.g., $a \cdot x \pm b = c$; $a \cdot x^2 + b^x = c$, and the like."[198] We can be interested in considering the inverse of a ternary function f as for example $f(a, x, b,) = a \cdot x + b$, i.e. the function $f^{-1}(a, b, c) =$ *the unique x, if it exists, such that* $f(a, x, b,) = c$. It is right in this sense that *the inversion procedure* is a particular case of that *more general form of inversion* which Husserl calls 'solution of equations'.

With this last method for generating new operations we have considered *all* the generation procedures identified by Husserl in order to circumscribe 'the totality of conceivable arithmetical operations'. And, in our judgment, this is indeed the most innovative idea to be found in the final chapter of the *Philosophy of Arithmetic*: as far as we know Husserl is the first scholar who, having insisted on the algorithmic meaning of arithmetical operations, explicitly specifies a number of general procedures by means of which new arithmetical (computable) operations are generated from given ones, and at the same time attempts to investigate the question concerning the characterization of the class of *computable* arithmetical functions *as a whole*. In Appendix 1 we defend the thesis that the generation procedures that Husserl studies in the 13th chapter of the *Philosophy of Arithmetic* give indeed rise to a class of numerical functions that is extensionally equivalent to the one known in contemporary logic as the class of 'partial recursive functions'.

The *Philosophy of Arithmetic* concludes with these words:

> The fact that in the overwhelming majority of cases we are restricted to *symbolic number formations* forces us to a rule governed elaboration of the number domain in the form of a *number system* ([...]) that according to a fixed principle always selects one from among the totality of the symbolic formations corresponding to each actual number concept and equivalent to it, and simultaneously assigns that one symbolic formation a systematic position. For every other conceivable number form there then arises the problem of evaluation: i.e. of classificatory reduction to the system number equivalent to it. But a survey of the conceivable forms of number formation taught us that the invention of appropriate methods of evaluation is dependent upon the elaboration of a *general arithmetic*, in the sense of a general theory of operations."[199]

1.17 Appendix 1: Husserl's Computable Functions[200]

As a result of our analysis of the 13th chapter (*The logical sources of Arithmetic*) of the *Philosophy of Arithmetic* we have a complete inventory of the various methods

[197] That is, the 'conditions' of the 'second sequence' above.

[198] *PoA.*, 298; *PdA* 282.

[199] *PoA* 299; *PdA* 283 (italics in the original).

[200] This appendix is excerpted from Centrone 2006.

for generating new operations which Husserl considers to try to dominate the 'totality of conceivable arithmetical operations'. Summing up, we have:

(i) Certain elementary or "initial" operations: addition and multiplication (which are *total*), and subtraction and division (which are *partial*)
(ii) Certain procedures by which new operations are generated out of other given operations, and specifically: (a) the schema of primitive recursion - which is not explicitly isolated, but rather the application of which is clearly exemplified in the construction of the 'Ackermann's succession'; (b) inversion – that is to pass from a function *f* to its inverse function or functions – viewed as a special case of the more general procedure of determining new numbers by means of equations; (c) composition, which consists in "putting together some functions (*Operationsmischungen*)"

It is now quite natural to ask whether it is possible to give a rigorous (mathematical) definition of Husserl's notion of the 'totality of conceivable operation in calculation', and moreover whether it is possible to prove that this class of functions, which we will call 'class *H*' (functions *à la Husserl*) corresponds, i.e. is extensionally equivalent, to the class of functions known, in computability theory, as the class of *partial recursive functions*.[201] To this aim, we think it advisable to review a number of important definitions.

A *n-ary* ($n \geq 1$) *partial numerical function* is a correspondence *f* which associates to each element of a certain subset *D(f)* of the set N^n of all the ordered *n*-tuples of natural numbers – the *domain* (of definition) of *f* – one and only one natural number. In the case that *D(f)* coincides with N^n, *f* is said to be *total*. When dealing with expressions, more precisely *terms*, which may be undefined (that is, expressions which may not denote anything, such as for instance, in the context under consideration, the expression *f*(2) when the argument 2 does not belong to the domain of *f*) the usual identity relation (denoted by '$=$') is conveniently replaced by the so-called "*Kleene* equality" relation (here denoted by the symbol '$\approx$'): $t \approx s$ holds if and only if the terms *t* and *s* are *either* both undefined, *or* are both defined and $t = s$.

A function *f* is said to be (intuitively) *computable* when there exists a mechanical procedure which, for every *n*-tuple **x** ($=x_1, \ldots, x_n$) of numbers belonging to *D(f)*, allows us to compute in a finite number of steps *f*(**x**), the value of *f* for the arguments **x**. As we saw in the preceding section, the fundamental problem of finding an adequate *mathematical* characterization (or *definition*) of the *intuitive* notion of computability – in other words, the problem of elaborating a *mathematical theory of the notion of effectiveness* (*in principle*) – has received various answers since

[201]This question was originally raised in Casari 1991: "It would be really worth to further investigate Husserl's attempt to dominate the *totality of all conceivable arithmetical operations*, as Husserl calls it. For, we also believe not to get wrong by saying that this is, most likely, the first characterization of the class of functions nowadays known as the class of *partial recursive functions*" (46). Indeed, our investigations concerning this issue, which finally led to the result presented here, originate from Casari's insightful suggestions to attempt at a mathematical reconstruction of Husserl's intuition.

1933–34 which, although conceptually distinct, have turned out to be extensionally equivalent. The approach considered below, which comes essentially from Gödel and Kleene, is best suited for our aims and is characterized by the fact that the class of functions which is proposed as a formal candidate capable of capturing the notion of partial computable function is defined *inductively*.

Definition. *The class* $\mathbf{P\mu}$ *of partial* μ*-recursive functions is the smallest class of partial numerical functions which*

(i) contains, as "initial" functions, the total unary functions Z (*constant-zero*: $Z(x) = 0$) and s (*successor*: $s(x) = x + 1$) and, for $k \geq 1$ and $1 \leq i \leq k$, the total k-ary functions $p_{k,i}$ (*projections*: $p_{k,i}(x_1,\ldots,x_k) = x_i$);

(ii) is closed under *substitution*, or *composition*: given a n-ary function h in $\mathbf{P\mu}$ and n k-ary functions $g_1, \ldots, g_n$ in $\mathbf{P\mu}$, the k-ary function f defined by:

$$f(x_1,\ldots,x_k) \approx h(g_1(x_1,\ldots,x_k),\ldots,g_n(x_1,\ldots,x_k))$$

belongs to $\mathbf{P\mu}$. We denote such a function f by $\mathsf{S}(h; g_1,\ldots, g_n)$;

(iii) is closed under *primitive recursion*: given a k-ary function g and a $k + 2$-ary function h in $\mathbf{P\mu}$, the unique $k + 1$-ary function f satisfying the two conditions:

$$f(x_1,\ldots,x_k,0) \approx g(x_1,\ldots,x_k)$$
$$f(x_1,\ldots,x_k,y+1) \approx h(x_1,\ldots,x_k,y,f(x_1,\ldots,x_k,y))$$

belongs to $\mathbf{P\mu}$. We denote such a function f by $\mathsf{R}(g, h)$;

(iv) is closed under *unbounded minimization*: given a $k + 1$-ary *total* function h in $\mathbf{P\mu}$, the k-ary function f such that:

$$f(x_1,\ldots,x_k) \approx \mu y(h(x_1,\ldots,x_k,y) = 0)$$

(where '$\mu x(\ldots x \ldots)$' denotes *the least* number n s.t. $\ldots n \ldots$, in case there exists a y s.t. $\ldots y \ldots$, and is undefined otherwise) belongs to $\mathbf{P\mu}$. We denote such a function f by $\mathsf{M}(h)$[202].

Concerning point (iv), note that, although M is required to take a *total* function h as argument, in general the function $\mathsf{M}(h)$ does not need to be total; actually, a necessary and sufficient condition for the totality of $\mathsf{M}(h)$ is that h satisfy the *regularity property*: $\forall \mathbf{x} \exists y\ (h(\mathbf{x}, y) = 0)$.

[202]Equivalently, the closure of the class $\mathbf{P\mu}$ under *unbounded minimization* may be formulated as follows: given a k+1-ary relation R which is *recursive* (that is, such that its associated *characteristic* function χ_R is in $\mathbf{P\mu}$), the k-ary function f such that $f(x_1, \ldots, x_k) \approx \mu y(R(x_1, \ldots, x_k, y))$ belongs to $\mathbf{P\mu}$.

Definition. *The class* $\boldsymbol{R\mu}$ *of total* μ*-recursive functions is the smallest class of total numerical functions which includes the initial functions of point (i) above, and is closed under the operators* S, R e M, *the latter applied to functions satisfying the regularity condition.*

Now, in Husserl's presentation of the totality of conceivable arithmetical functions we also find certain *basic*, *initial functions* (the four *elementary* operations), and *certain procedures by means of which new operations can be generated*; in particular, among the above mentioned ones, the composition procedure (*Operationsmischungen*), which clearly is adequately reflected on the formal level by the operator S (given that projections are present), and – although not so explicitly, as we already observed – primitive recursion R. Minimization (M) is not considered by Husserl, but on the other hand the *inversion* procedure is explicitly singled out. It is worth noticing here that the latter generative procedure – more precisely, a restricted form thereof – also plays a central role in the interesting equivalent characterization of the class of *total* recursive functions as presented by Julia Robinson,[203] namely:

Definition. $\boldsymbol{JR'}$ *is inductively defined as the smallest class of unary, total numerical functions which*

(i) contains, as initial elements, the functions s (successor) and E (*excess over a square*): $E(x) = x - n^2$, where n is the *greatest number* such that $n^2 \leq x$ *and* $(n + 1)^2 > x$;

(ii) is closed under

- *restricted composition*: for any two functions f and g in $\mathbf{JR}'$, the function h such that $h(x) = f(g(x))$ (i.e.: $\mathrm{h} = \mathsf{S}(f;g)$) belongs to $\mathbf{JR}'$;
- *addition of functions*: for any two functions f and g in $\mathbf{JR}'$, the function h such that $h(x) = f(x) + g(x)$ (i.e.: $h = \mathsf{S}(+; f,g)$) belongs to $\mathbf{JR}'$;
- *inversion of surjective functions*: for every function f in $\mathbf{JR}'$ such that $\forall n \exists m$ $(f(m)=n)$, the inverse-function of f, defined as the function h such that $h(x) = \mu y(f(y) = x)$, belongs to $\mathbf{JR}'$. We will denote this function by $\mathrm{I}(f)$.

We have to point out that the reason why we use the operator μ is that f is not necessarily injective: so, among the many possible arguments y s.t. $f(y) = x$ (at least one does exist, since f is surjective), we choose the smallest.

Definition. *Let* J *be the binary function ('pairing function') such that*

$$J(x,y) = [(x+y)^2 + 3x + y]/2$$

[203] Robinson 1950. For a clear presentation, see also Yasuhara 1971, 110–117.

It is not difficult to prove that J belongs to $\mathbf{R\mu}$ (actually, J is *primitive recursive*)[204] and that J puts into a *one-to-one correspondence* the set N^2 of all the ordered pairs of natural numbers with the set N of all natural numbers.

Theorem. *The class* $\boldsymbol{R\mu}$ *of all total recursive functions coincides with the closure under substitution of the class* $\boldsymbol{JR'}$ *extended with the function* J*; that is to say: every total recursive function can be obtained by repeated application of the operator* S*, starting from (unary) functions in* $\boldsymbol{JR'}$***, and*** $\boldsymbol{J}$.

Concerning the characterization of $\boldsymbol{R\mu}$ through inversion which follows from the above result, it is important to notice that

(i) only *unary* functions are considered and inverted;
(ii) moreover, only unary *surjective* functions are inverted (surjectivity obviously being a necessary condition to have totality).

In contrast, in the 13th chapter of the *Philosophy of Arithmetic* (at least, in the examples given by Husserl) inversion of functions which are, in general, *neither* surjective *nor* unary are considered. Two consequences follow:

First, given a unary, possibly non-surjective function f, its inverse function $\mathsf{I}(f)$ may be *partial*: $\mathsf{I}(f)(m)$ is undefined, whenever m is s.t. there does not exist any number n satisfying $f(n) = m$.

Next, let us consider a *binary* function f. In what way can one speak of the inverse function of f? Normally, when one refers to numerical functions one means functions which associate *numbers*, and not pairs or triples or ... of numbers, to numbers: in other words, values are always elements of N, and not of $N^2, N^3, \ldots$ Husserl's solution consists in the association of *two* inverses to such a f. For instance, if f is the *exponential function*, its first inverse function is the *x-th root of y* function:

$$Rad(x, y) \approx \text{the unique (if it exists) positive integer } z, \text{ s.t.} z^x = y$$

While its second inverse function is the *base-x logarithm of y* function:

$$Log(x, y) \approx \text{the unique (if it exists) positive integer } z, \text{ s.t.} x^z = y.$$

So, keeping to the binary case – the extension to the n-place case with n greater than 2 being straightforward, as we'll see – we will assume, in full generality,[205]

---

[204] A total function f is *primitive recursive* if it can be obtained from the initial functions Z, s and $p_{n,i}$ by means of the operators S e R. In the expression defining J, note that the numerator is always an even number, so that dividing by 2 makes sense.

[205] Notice that – the assumption that f is injective in both places separately, like the exponential function, being too restrictive – we have replaced the operator ι, 'the unique ... such that ...', with the operator μ, 'the minimum ... such that ...'. We will come back later on this rather delicate point.

that every function f is associated with the two inverse functions, both possibly partial, $\mathsf{I}_1(f)$ and $\mathsf{I}_2(f)$:

$$\mathsf{I}_1(f)(x,y) \approx \mu z(f(z,x) = y$$
$$\mathsf{I}_2(f)(x,y) \approx \mu z(f(x,z) = y$$

Given the above preliminary clarifications, we can now readily pass to the presentation of our proposal of a rigorous formal definition of the class **H** of arithmetical functions *à la Husserl*, justifying our choices in the concluding paragraphs.

Definition. ***H*** *is the class which is defined inductively (analogously to the classes* $\boldsymbol{P\mu}$, $\boldsymbol{R\mu}$ *and* $\boldsymbol{JR'}$*) as the smallest class of partial numerical functions which*

(i) contains the initial functions: $+$ (*addition*), $\cdot$ (*multiplication*), $-$ (*subtraction*, partial), : (*division*, partial), plus the projections $p_{k,i}$;
(ii) is closed under the following generating procedures:

- *substitution* (*Mischung*) and *primitive recursion* (operators S, resp. R), as for $\mathbf{P\mu}$;
- *inversion*, defined as follows: given a *total* $n+1$-ary ($n \geq 0$) function f in **H**, the $n+1$-ary functions $\mathsf{I}_1(f), \ldots, \mathsf{I}_{n+1}(f)$, such that (for $1 \leq i \leq n$)

$$\mathsf{I}_i(f)(x_1, \ldots x_n, y) \approx \mu z(f(x_1, \ldots x_{i-1}, z, x_i, \ldots x_n) = y)$$

belong to **H**.

Now, it is not difficult to prove that **H** contains the functions *constant-zero* and *successor* and is closed under the operator M of *unbounded minimization.*

First of all, since clearly $x - x = 0$ and $x + (x : x) = x + 1$ for every x, we have $Z(x) = x - x$ and $s(x) = x + (x : x)$; therefore
$Z \equiv \mathsf{S}(-; p_{1,1}, p_{1,1})$ and $s \equiv \mathsf{S}(+; p_{1,1}, \mathsf{S}(:; p_{1,1}, p_{1,1}))$ belong to **H**.

Next, let h be a *total* $k+1$-ary function in **H**. Since **H** is closed under *inversion of total functions,* **H** contains – in particular – the $k+1$-th inverse of h, that is the function $\mathsf{I}_{k+1}(h)$ such that (with $\mathbf{x} \equiv x_1, \ldots x_k$):

$$\mathsf{I}_{k+1}(h)(\mathbf{x}, y) \approx \mu z(h(\mathbf{x}, z) = y)$$

Let us now consider the k-ary function f defined by:

$$f \equiv \mathsf{S}(\mathrm{I}_{k+1}(h); p_{k,1}, \ldots, p_{k,k}, \mathsf{S}(Z; \mathrm{p}_{k,1}))$$

Since **H** is closed under *substitution* (S) and contains the functions Z (as we have already verified) and the projections (which are initial functions), it holds that f belongs to **H**.

But

$$\mathrm{f}(\mathbf{x}) \approx \mathrm{I}_{k+1}(h)(p_{k,1}(\mathbf{x}),\ldots,p_{k,k}(\mathbf{x}),\mathsf{S}(\mathrm{Z};\mathrm{p}_{k,1})(\mathbf{x})) \approx \mathrm{I}_{k+1}(h)(x_1,\ldots,x_k,\mathrm{Z}(\mathrm{p}_{k,1}(\mathbf{x})))$$
$$\approx \mathrm{I}_{k+1}(h)(x_1,\ldots,x_k,\mathrm{Z}(x_1) \approx \mathrm{I}_{k+1}(h)(x_1,\ldots,x_k,0) \approx \mu\mathrm{z}(\mathrm{h}(x_1,\ldots,x_k,\mathrm{z})=0$$

and so $\mathsf{M}(h) \equiv f$ belongs to **H**.

Incidentally, we observe that if we add to the initial functions both the characteristic function of identity ($\varepsilon(x, y) = 0$ if $x = y$, and $= 1$ otherwise) and of the strict order relation ($m(x,y) = 0$ if $x < y$, and $= 1$ otherwise), then – given that addition and multiplication are initial functions – we may even dispense of requiring closure under the operator R of primitive recursion.[206]

As an immediate corollary of what has been established above, we find that the classes **H** and **Pμ** are extensionally equivalent, and so that the class of all 'conceivable arithmetical functions' isolated by Husserl coincides – *assuming* the adequacy of our formal reconstruction – with the class of all partial numerical *computable* functions (modulo the *Turing-Church Thesis*).

It might be objected that our reconstruction is, at least in certain respects, *ad hoc*. To begin with, why should one add projection functions to Husserl's initial operations (the four elementary operations)? This objection can be easily met. The inclusion of the projections among the initial functions is dictated exclusively by the necessity to cope with the "rigidity" of the operator S, by means of which alone it is clearly impossible to construct obvious 'compositions' in which e.g. one identifies some variables, as in some of the *Mischungen* which Husserl himself considers (for instance, $x^2 + \mathrm{b}^x$). A second, and according to us more serious, objection, might point to the fact that, in order to generalize the examples of *inversion* explicitly considered by Husserl and to introduce an inversion operator (the operator I) – with the aim of obtaining a class of functions extensionally equivalent to **Pμ** –, we have made use, in the definition schema of the latter, of the operator μ. In other words: μ is already an ingredient of our general operator of inversion. Now, on the one hand this circumstance is not automatically tantamount to saying that **H** is closed under *unlimited minimization* (in other words: this has in any case to be proved). On the other hand we have to admit that closure of **H** under *unlimited minimization* can be easily expected – and in fact the proof given above is extremely simple.

With regard to this second objection, it is our present opinion that the possibility of a suitable weakening of the operator I, which is still capable of yielding closure under M without containing minimization as an explicit ingredient, cannot be excluded.

---

[206]The reason being the following: **H**, modified as indicated, turns out to be still closed under minimization and to contain a β-function, and these features are sufficient to express primitive recursion.

In brief, we think the crux of the question is as follows. As we have seen, all the examples which Husserl gives of the inversion procedure concern binary functions h which – like *exponentiation* – are *injective* in both the arguments, that is, satisfy the conditions:

$$h(x,z) = h(y,z) \rightarrow x = y \quad \text{and} \quad h(z,x){=}h(z,y) \rightarrow x = y.$$

In this case, in order to define the two inverse functions $\mathsf{I}_1(h)$ and $\mathsf{I}_2(h)$ we may equivalently use, in place of the operator μ, the operator $\imath$ ('the unique number, if it exists, such that ...'):[207]

$$\mathsf{I}_1(h)(x,y) \approx \imath z(h(z,x) = y$$
$$\mathsf{I}_2(h)(x,y) \approx \imath z(h(x,z) = y.$$

We do not know whether the restriction to the applicability of the operator I, defined via $\imath$ as above, to injective functions, is such that closure under minimization, hence the equivalence of the class **H** modified in this way with the class **Pμ** of partial recursive functions,[208] is still provable. What is certain is that, in the chapters of the *Philosophy of Arithmetic* we have analyzed, there is no example of inversion of *non-injective* functions, yet one cannot find clear elements to support the conjecture that by inversion Husserl meant inversion of injective functions *only*. So, to conclude, the closure under *inversion* which Husserl has in mind might be *weaker* than the one we postulated in the definition of the class **H**.

1.18 Appendix 2: On Operations, Algorithmic Systems, and Computation

'Operation', 'computation', 'algorithm' are, as we have seen, fundamental and interrelated key-notions that enter into many of Husserl's reflections in his *Philosophy of Arithmetic*. It will be rewarding to have a close look at two short treatises from the *Nachlass* that have been sadly neglected in the literature, as they will help us to throw further light on this group of concepts.

[207]Note that, in case there is more than one z such that $P(z)$ holds, $\imath zP(z)$ turns out to be undefined whereas $\mu zP(z)$ is defined. For instance, let $h(0, x) = 3$ and $h(1, x) = 3$: then $\imath z(h(z, x) = 3)$ is *undefined* and $\mu z(h(z, x) = 3)$ is defined and equal to 0.

[208]While it is clear that inversion I, once it is defined with $\imath$ in place of μ and not restricted to injective functions, doesn't preserve computability.

1.18.1 On the Concept of the Operation

Husserl's treatise "*Zum Begriff der Operation*"[209] derives from a manuscript that contains studies for the planned second volume of the *Philosophy of Arithmetic*. Evidence for the editor's conjecture that the text dates from 1891 is the fact that Husserl scribbled '18.XI.91' on a page of the manuscript.[210] On the manuscript cover he wrote:

> Concept of "combination" ("operation" within a *Mathesis*). Basic operation. Partition of combinations. — Detailed investigation: why equivalent combinations ([i.e.] operations that evidently "come to the same thing") are regarded as the same operation whereas in other cases equivalences are affirmed as valuable propositions (*wertvolle Sätze*)? Which equivalences are [to be] affirmed as propositions within a *Mathesis*, and which ones must be regarded only as different expressions of the "same" proposition?[211]

The text, far from offering a unified and systematic account of the above topics, deals in a rather fragmentary and tentative style with a number of extremely interesting problems and conceptualizations having to do with the role of 'equivalences (*Äquivalenzen*)' in deductive theories and algorithmic systems, as well as with the notions of *combination* (*Verknüpfung*) and *operation* (*Operation*). For the sake of exposition and analysis it is convenient to start with the latter issue.

1.18.1.1 Combinations and Operations

Husserl uses the term 'combination (*Verknüpfung*)' to mean, in full generality, any kind of *conceptual synthesis* of two or more objects which determines a new object called the 'result (*das Resultat*)' of the combination. Hence we find combinations not only between numbers, that is, within the arithmetical domain, but between objects of any domains.

> Wherever a conceptual determination is present which determines one object by means of other objects, we speak of a combination (synthesis) of the latter objects into the former, the "result of the combination." Thus, for example, we call any kind of conceptual production (*Herstellung*) of one number from two or more numbers a combination (additive, multiplicative, etc.) of those numbers.[212]

The objects entering in a combination either all belong to the same class, in which case a new object belonging to the same class is determined, or each of them belongs to a different class, in which case the object determined by the combination bears the conceptual determinations of each of the classes to which the combined objects belong.

---

[209] *PdA* App. 408–429; *PoA* 385–408. Cp. Centrone 2005 for the following discussion.

[210] *PdA* App. *Textkritische Anmerkungen* 538.

[211] Loc. cit.

[212] *PdA* App. 422; *PoA* 400.

Husserl often calls the objects which are connected in a combination 'members (*Glieder*)' of that combination. Since in current algebraic jargon the word 'combination' (in German, *Verknüpfung*) is sometimes used as a synonym of '(binary) total function over a given set', it is important to stress that this is not at all the intended meaning in Husserl's treatise. It clearly emerges from what he says that the *combination* is not at all what effects the connection between its members – the *operation*, as we would say – but rather the complex constituted by objects which are connected (the members) *plus* their combination – the *explicit presentation* of the result, as we would put it nowadays. So in the case of $a + b$, for instance, the combination *is not* the operation of addition but the very complex $a + b$. To put it in another way, from the formal-morphological viewpoint of current logic Husserl's 'combinations' correspond to *complex individual terms* of an elementary language, that is, to those complex expressions which are built from individual variables and constants by means of iterated application of function letters.

The *operation* is rather the arithmetical counterpart of the 'combining thought (*verknüpfender Gedanke*)' that is embedded in a combination. In the treatise we find some interesting though rather sketchy notes on the nature of operations and their connection with the notion of *production* (*Erzeugung*) of an object. Husserl also briefly considers the question whether one really needs *at least two objects* to have a combination, a requirement he had indeed presupposed throughout the preceding pages. He takes as an example what happens with *negation* in the 'logical calculus' (read 'calculus of classes'), where to each class a there is associated its complementary class (a', in Husserl's notation). His proposal is either to treat 'unary' operations as binary univocal relations ("We can interpret the situation as involving a relation: $a \ \phi \ b$. Then b is to be a'; a' unambiguously determinate"), or to generalize the notion of operation itself: "An operation is a way of deriving new numbers from one or several numbers; or, from one object: negation, inversion, coincidence".[213]

A certain ambiguity in Husserl's talk of operation should also be mentioned: he employs the term 'operation' both to designate a certain subjective activity which can be performed on aggregates, that is, uniting (*Zusammenfügung*) and dividing (*Teilung*) and to designate the 'forms of numerical determination (*Formen der Zahlbestimmung*)', that is, the arithmetical operations. It is hard to see how a subjective activity could be a mathematical object satisfying certain general laws. One stumbles here over the difficulty Husserl himself describes in the *Preface* to the first edition of the *Prolegomena*: "I became more and more disquieted by doubts of principle, as to how to reconcile the objectivity of mathematics, and of all science in general, with a psychological foundation of logic."[214]

We stick here to the second use of the term 'operation': under this reading, to repeat, operations are 'forms of numerical determination', forms of determining new numbers by means of given numbers. These forms are founded upon (or, in a

---

[213] *PdA* App. 427; *PoA* 405.

[214] *PR* VII, *PRe* 42.

sense, generated by) subjective activities such as *Addieren* and *Dividieren*. According to Husserl, "on such concepts of determination all the law-like regularities are grounded that prevail in the domain of number and that are to be investigated by arithmetic".[215] Obviously, not all possible forms of numerical determination are of interest to arithmetic. It considers only determinations of numbers *by means of numbers*, hence it is not concerned at all with determinations like "the number of flowers in this garden".[216]

Husserl gives the following outline of the manner in which arithmetic must proceed in order to obtain general laws concerning these 'forms':

(i) In order to discover the most general laws of the numerical domain arithmetic has to leave any *specific* determination of numbers out of consideration.
(ii) Therefore, each number has to be regarded only as "a certain number (*eine gewisse Zahl*)", as "an *arbitrary* (*irgendwelche*) number". This is the reason why in general arithmetics numerals are replaced by letters *a*, *b*, etc.[217]
(iii) Once this preliminary abstraction has been effected, it is to be asked in which forms new numbers can be determined from arbitrarily given numbers,
(iv) and as soon as these 'forms of construction' are found, the task is to specify the general laws to which they are subject. In Husserl's own words, "when we have found such forms, we think of 'any (*irgendwelche*)' numbers as united by them, and then we ask ourselves which general laws result from the concepts of these forms of construction."[218]

Steps (ii) and (iii) rest on the possibility to pass from a concretely given combination to a 'form of combination'. This form is obtained by replacing the objects (that is, the members of the combination) by *symbols for objects,* and by replacing the combination (i.e. the operation) by a *symbol for an operation.* For instance, to get the form of the concrete combination $5 + 7$ we must first of all substitute letters for the determinate numerals '5' and '7'. In this way, each of the two members of the given combination is thought of only as "a certain number (*eine gewisse Zahl*)", and now one just has to stipulate that, inside a formal expression, different occurrences of the same letter always refer to the same object (and, of course, that different objects are referred to by occurrences of different symbols). Yet the form we have so far obtained, $a + b$, is still bound to a specific domain, namely that of natural numbers. To build up a more abstract formal theory (a general arithmetic), one has to go one step further and pass to the *general form* of that combination in which the addition sign is replaced by an indeterminate *operation symbol*, for instance ρ. So the general form of the concrete combination we started from is $a \rho b$. As in the case of the letters a and b one now stipulates that inside an expression *as well as* inside a 'formal theory' different occurrences of the

[215] *PdA* App. 408; *PoA* 385.
[216] Loc. cit.
[217] *PdA* App. 409; *PoA* 386.
[218] *PdA* App. 408; *PoA* 385.

same operation symbol always denote the same operation (and, of course, that distinct operations are denoted by different symbols).[219]

In Husserl's treatise the idea of arriving at the form of a combination by enucleating, as it were, a concretely determined combination at different levels of abstraction is elaborated at great length. Although his reflections are neither always accompanied by precise definitions nor always entirely coherent, it is worth trying to reconstruct some of them – as an impressive sample of his interest in the algebraic way of thinking.

Both for a concretely determined combination and for a general form of combination Husserl distinguishes material types from material modes – and formal types from formal modes. The '*material type* (*der sachliche oder innere Typus*)' of a combination whose members are, say, numbers is the kind of connection (e.g. additive, multiplicative, mixed, etc.) that combines the members, that is, *der verknüpfende Gedanke*. For example, $a + b$ and $b + a$ belong to the same material type, and so do $(a + b) + c$ and $(a + c) + b$, whereas $a + b$ and $(a + b) + c$ do not belong to the same material type, although they share the kind of combination (*Verknüpfungsart*, see below). In a type the number of the members is relevant but not their order.

As regards the '*material mode* (*der materiale Modus*)', the position of the members inside a combination does matter: given a type there are as many material modes corresponding to that type as there are different ways to permute the order of the same members.[220]

> If in a non-symmetrical combination we interchange the members occupying different positions, then the "mode of combination" of the members changes, and thus, in a certain manner, the concept of that combination as well. Nevertheless, the general concept of combination – the type and the kind of the combination – remains unaltered. Consider, for example, the combination $a + b$. In $b + a$ the same members are combined in a different way, but the type is the same: the addition of a number to another.[221]

All general laws concerning combinations, Husserl says, concern 'modally' determinate combinations.[222]

The '*formal type* (*der formale oder äußere, reine Typus*)' of a combination is characterized as follows:

> By the formal type of a combination we understand the concept that results from its type through the following abstraction: As regards its members we also abstract from the fact that they are objects of domain D, thus retaining each of them merely as an object in general. On the other hand, we also abstract from the specific nature of the elementary combinations constituting the combination as a whole – from combinations of the same kind merely retaining merely the fact that they are combinations of the same kind, and from combinations of different kinds retaining only the fact that they are combinations of different kinds. [223]

[219] *PdA* App. 427; *PoA* 404.
[220] *PdA* App. 420; *PoA* 397–398.
[221] *PdA* App. 423; *PoA* 401.
[222] *PdA* App. 425; *PoA* 402.
[223] *PdA* App. 426; *PoA* 404.

The '*formal mode* (*der formale Modus*)' – basically a specific pattern of positions of the members in a completely abstract form of combination, i.e. in a formal type – results from a material mode by means of an abstraction that is parallel to the one leading from a material to a formal type. "If [in $(a\ \rho\ b)\ \rho'\ c$] we switch the letters for the members, arriving at e.g. $(b\ \rho\ a)\ \rho'\ c$, we obtain another mode, but the formal type is the same."[224]

Apart from the type and the mode of a combination, Husserl also introduces the notion of a "kind of combination (*Verknüpfungsart*)". At a first sight, at least, this seems to be the main operation of a combination. Actually, Husserl gives two different definitions of this concept. According to the first one, the *Verknüpfungsart* is the connecting thought of a *homogeneous* combination of two or more members:

> If we direct our attention entirely away from the members and attend exclusively to the thought that brings about their synthesis, then we obtain the kind of combination. We speak, for example, of addition, subtraction, etc., as different kinds of number-combinations; and wherever in so doing we make clear to ourselves the concept of addition in *concreto*, we obviously pay no attention at all to the individual members of the particular addition that serves as our basis. Hence for the concept of the kind of combination the number of the members is non-essential, whereas the type is immediately modified with any modification of the number of the members.[225]

In this sense the combinations $a + b$ and $(a + b) + c$ have the same *Verknüpfungsart* but they are not of the same type, since, as we saw, the type varies with the number of the members of the combination.

As for the second definition of *Verknüpfungsart*, we must invoke the Husserl's distinction between 'simple (*einfache*)' and 'complex (*zusammengesetzte*) combinations'.[226] A combination is said to be simple when its members are not themselves combinations, as in cases like $a + b$ and $a \cdot b$, and a combination is said to be complex otherwise, as in cases like $a \cdot (b + c)$ and $(a + b) + (a \cdot b)$. Appealing to this distinction Husserl says that "the type of simple combinations is called the *Verknüpfungsart*".[227] So according to this definition (as opposed to the first one) a *Verknüpfungsart* comes with a determinate number of argument-places, a determinate *-arity*, as we would nowadays put it. Under the second reading of the term '*Verknüpfungsart*' – which is the one Husserl actually employs in the course of his investigations – we can simply identify a *Verknüpfungsart* with an *operation* in the current mathematical sense.

Having fixed the meaning of 'operation' we can now move on to a related notion that will play an important role in some of Husserl's reflections on "equivalences" and algorithmic systems which are to be considered in the next sub-section.

Given the distinction between simple and complex combinations, Husserl identifies in the processes of instantiation (*Besonderung*) and composition

---

[224] *PdA* App. 427; *PoA* 404.

[225] *PdA* App. 425; *PoA* 403.

[226] *PdA* App. 426; *PoA* 404.

[227] *PdA* App. 420; *PoA* 397.

(*Komposition*) the two fundamental ways out of which combinations are generated. By means of *instantiation* we generate e.g. the combinations $a + b, b + a, a + c$ etc. as particularizations of the additive combination. By means of *composition* we obtain a new (form of) combination out of a given (form of) combination by replacing one of its members by another (form of) combination, not necessarily one of the same type as the one we started from. For example, from $a + b$ we may obtain $a + (b + c)$ through composition, and then in turn, e.g., $(a \cdot d) + (b + c)$, and so on.

Given several operations $V_1, \ldots, V_n$, Husserl calls the potentially infinite totality consisting of all combinations that can be constructed, by means of instantiation and composition, from those operations and the associated simple types the '*sphere* (*Bereich*)' of combinations associated with those operations.

> ... we understand by the *sphere* of certain kinds combination (*Verknüpfungsarten*) the totality of the determinate combinations which fall under those kinds or arise by composition from combinations belonging to them.[228]

Thus, for instance, the sphere of $(+, \cdot)$ contains the combinations $a + b, b + a, a + (b \cdot c), a + (b \cdot a), (a \cdot a) + (b \cdot (a + b)), \ldots$ Formally speaking, we can simply identify the sphere of $V_1, \ldots, V_n$ with the set of all terms that can be generated starting from variables and, possibly, individual constants, by means of the function symbols $V_1, \ldots, V_n$.

1.18.1.2 Equivalences and Algorithmic-Deductive Systems

When Husserl investigates in his treatise "*On the concept of the operation*" the role of 'equivalences (*Äquivalenzen*)' and of 'transformations into immediate equivalents' in the deductive disciplines, his main question can be rephrased in modern terminology as follows: how is the abstract-formal approach set up with respect to typical (arithmetical or more abstract) structures constituted by a domain D plus a number of operations and (possibly) relations on D to be specified? This main question is addressed by Husserl *via* three interrelated sub-questions:

(a) Which cognitions (*Erkenntnisse*)[229] are from a logical, (algebraic) deductive standpoint *relevant* as regards a certain structure, and which ones are not?
(b) Which kinds of *equivalent* cognitions are to be explicitly expressed in a formal theory?
(c) How is the algorithm, that is, the formal deductive theory finally set up?

Let us try to extract, from an extremely unsystematic collection of claims and observations, what seem to be the main points Husserl is driving at.

---

[228] *PdA App. 419; PoA 396.*

[229] *Faute de mieux* I use 'cognitions' as my rendering of '*Erkenntnisse* (propositions that have become contents of knowledge)'.

As we saw, Husserl has it that arithmetic considers only *general* forms of determination/construction of numbers from numbers, and their laws. The keyword here is '*general*'. The determination of new numbers has to be performed with utmost generality, and the laws that come into consideration for arithmetic are *general* laws of the form 'All numbers have such-and-such a property'. For example, addition is a general, uniform method of constructing a number $a + b$ once two arbitrary numbers a and b are given, and e.g. *commutativity* and *associativity* are appropriate laws of this particular form of determination of numbers.

> The determining numbers remain necessarily indeterminate in the course of these investigations. They are only ... thought of as in some way determinate or determinable. Therefore the numbers signs which are used here are not numerals, but rather are arbitrarily chosen letters.[230]

Given the implicit algebraic framework underlying these reflections, the general laws Husserl is here thinking of have the form of (tacitly universally quantified) *equations*, that is of *equivalences between concepts* (see below). But is every kind of valid equivalence to be taken into account when an algorithm is set up? Husserl asks us to consider the following three concepts of operation, which are only psychologically distinct: 'the result of the union of a and b', 'the result of the attachment of a to b', 'the whole which is decomposable into a and b'. There is, *from a logical point of view*, no reason to distinguish them and to establish within a formal system an equivalence between them.

> The psychological difference between the concepts of combination in question is not yet a reason for distinguishing them logically. Generally, for the purpose of knowledge acquisition what is psychologically distinct can [logically] be fully equivalent.[231]

All this holds, however, with some essential restrictions. Husserl actually acknowledges that the logical separation of equivalents, as well as the transformation of cognitions into immediately equivalent cognitions can sometimes serve the purpose of *extending* our knowledge. This happens whenever a proposition – through "immediately equivalent transformation (*unmittelbar äquivalente Transformation*)"[232] – receives a form which makes it appropriate as a premise for certain inferences.[233] As an example Husserl mentions the 'immediate inferences' of traditional logic,[234] that is, the so-called conversions like **SiP** → **PiS** and **SeP** → **PeS**. Thus, for instance, given the propositions **MaP** and **MiS**, once the conversion from **MiS** to **SiM** is made, it becomes obvious that **SiM**, together with **MaP** as first premise, is the appropriate second premise for applying the syllogism in *Darii* that yields the conclusion **SiP**.

---

[230] *PdA* App. 408; *PoA* 385.

[231] *PdA* App. 411; *PoA* 388.

[232] *PdA* App. 413; *PoA* 390.

[233] As to Husserl's intellectual heritage in this respect, see the following section.

[234] Loc. cit.

Answering questions (a)–(c) above, Husserl attempts, first of all, to provide a taxonomy of all possible types of equivalences that are to be expressed in a formal system. He considers, substantially, equivalences of two kinds:

1. Equivalences among concepts
2. Equivalences among judgements

> Conceptual equivalences open up the possibility of discerning, through a sequence of equivalent transformations, equivalences of complicated determinations that are extremely remote and not immediately evident at all. The equivalences between judgements either function as forms of inference or supply general major premises for inferences to be carried out, and serve, apart from equational inferences, in the advancement of knowledge beyond any domain of equivalence.[235]

Generally speaking, "we learn from the foregoing considerations of what type the equivalences are that have an essential function in the deductive disciplines."[236] They are

(i) "*general logical*" *equivalences between judgements*, such as $a = b \leftrightarrow b = a$ (symmetry of equality) or a > b ↔ b < a (in general: Rab ↔ Rˇba). Such equivalences express laws – like those concerning equality and relations R with their converses Rˇ – which hold in every domain whatsoever. (Rxy ↔ Rˇyx is indeed a logical law of the operator ˇ which produces the converse of a binary relation.)

(ii) "*laws that belong the axioms or follow from them,*"[237] i.e. equivalences which depend on the particular domain axiomatized by the theory. They divide in turn into

   1. *Equivalences between judgements,* such as a = b ↔ a + c = b + c (in the numerical domain). Such equivalences are taken by Husserl as having the role of inference rules. In our example: from $a = b$ to infer $a + c = b +$ c, and conversely.
   2. *Equivalences between concepts:* they are equivalences among particularizations of some combination-type, such as that between $x + y$ and $y + x$ as a particularization of the additive type of combination.

Now suppose we have a structure S constituted by a certain domain D, certain operations $V_1, \ldots, V_n$ over D, and possibly certain relations. Which conditions should an acceptable algorithm, a deductive theory AX[S] for S satisfy? The answer follows by considering the 'cognitions' pertaining to this structure, which according to Husserl should be divided into the following groups:

---

[235] *PdA* App. 419; *PoA* 396.

[236] *PdA* App. 418; *PoA* 395.

[237] Loc. cit.

(i) "Cognitions (*Erkenntnisse*) which are founded upon a single concept of combination,"[238] that is, general laws which are valid for one specific operation applied to arbitrary terms, such as

$$x + y = y + x;$$
$$x + (y + z) = x + (y + z).$$

These equivalences are to be expressed in the corresponding axiom system AX[S] as specific axioms.

(ii) "Cognitions which are simultaneously founded upon several concepts of combinations,"[239] that is, founded on more than one operation. E.g.

$$x \cdot (y + z) = (x \cdot y) + (x \cdot z).$$

Husserl distinguishes three different sub-cases of (ii):

1. "The respective group of concepts of combination contains only concepts that are non-equivalent to each other,"[240] that is to say, the operations $V_1, \ldots, V_n$ are pairwise not equivalent. In this case we shall have in AX[S] laws of two types:

$$\begin{array}{ll} \textit{equations}, \text{such as} & \mathrm{a} \cdot (\mathrm{b} + \mathrm{c}) = (\mathrm{a} \cdot \mathrm{b}) + (\mathrm{a} \cdot \mathrm{c}); \\ \textit{inferences}, \text{such as} & \textit{from}\ \mathrm{a} > \mathrm{b}\ \textit{and}\ \mathrm{b} > \mathrm{c}\ \textit{to infer}\ \mathrm{a} > \mathrm{c} \end{array}$$

2. "The group of concepts of combination concerned contains only those that are equivalent to one other,"[241] that is to say, all the operations $V_1, \ldots, V_n$ are pairwise equivalent. In such cases it suffices to limit the system AX[S] *to a single, arbitrarily chosen operation of this group.*
3. "The group concerned contains combinations some of which are equivalent to one another and some of which are not."[242] Suppose $V_1, \ldots, V_i$ are pairwise non-equivalent, while $V_{i+1}, \ldots, V_n$ are pairwise equivalent. Then the $V_{i+1}, \ldots, V_n$ have to be reduced to a single operation V of the group. In other words, we have to shift from the consideration of the structure $\langle D, V_1, \ldots, V_i, V_{i+1}, \ldots, V_n \rangle$ to the consideration of the sub-structure $\langle D, V_1, \ldots V_i, V \rangle$.

Nothing essential from a logical point of view – Husserl stresses – *is lost in the above reduction.* It is in fact evident that "the totality of truths that are founded upon

[238] *PdA* App. 414; *PoA* 391.
[239] *PdA* App. 415; *PoA* 392.
[240] Loc. cit.
[241] *PdA* App. 415–416; *PoA* 393.
[242] *PdA* App. 416–417; *PoA* 394.

all of the types of elementary combinations $V_1, \ldots, V_n$ of a domain D undergo no diminution worthy of attention if we restrict ourselves to those combinations that are not equivalent to one another, and thus retain one from each group of equivalent combinations".[243] Husserl concludes:

> We therefore establish the following rule, valid for all domains of deduction: that among the elementary types of combination, equivalents are not to be tolerated. From each totality of equivalent, elementary types of combination, a single one is to be selected; all the remaining ones are to be ignored.[244]

Using the present-day algebraic terminology, we can say that all this simply amounts to the following contention: If $V_1, \ldots, V_n$ are the operations on D under consideration, no equational pieces of information are lost if we pass to the *quotient structure* of $\langle$D, $V_1, \ldots, V_n\rangle$ *modulo* the equivalence relation $\approx$ of extensional equality between operations.

It is worth mentioning that Husserl takes also into account further concepts and problems, all of a "metalogical" character, concerning more general kinds of reducibility of one operation to others. For instance:

(i) *Reducibility*: A type of combination V is said to be "reducible to the combination types $V_1, V_2, \ldots, V_k$" if among the latter there is one which is equivalent to V.[245] That is to say, V is reducible to $\{V_1, V_2, \ldots, V_k\}$ if and only if V is equivalent to a certain combination belonging to the sphere of $\{V_1, \ldots, V_k\}$. For instance, *double-of* (x) is reducible to the sphere of $\{+\}$, since *double-of* $(x) = x + x$.
(ii) *Group of irreducible combinations*: "A group of combination types is said to be irreducible if none of them is reducible to the others."[246] That is to say, $\{V_1, \ldots, V_k\}$ is irreducible iff for each i $(1 \leq i \leq k)$ V_i is not reducible to $\{V_1, \ldots, V_{i-1}, V_{i+1}, \ldots, V_k\}$.
(iii) *Group of basic combinations*: "An irreducible group to which all kinds of combination of a domain are reducible is called *a group of basic combinations*."[247]
(iv) *Problem*:[248] Suppose C and C′ are two equivalent (interreducible) groups of operations, respectively axiomatized by AX[C] and AX[C′], and A is a statement relating to C. Is it the case that from the fact that A follows from AX[C] & AX[C′] we can infer that A *already follows from* AX[C]?

---

[243] *PdA* App. 417; *PoA* 394.
[244] *PdA* App. 417; *PoA* 395.
[245] *PdA* App. 419; *PoA* 396.
[246] Loc. cit.
[247] *PdA* App. 419; *PoA* 397.
[248] *PdA* App. 416 *PoA* 393.

1.18.1.3 A Historical Postlude on Leibniz

In the introduction to his *Mechanism, Mentalism and Metamathematics* Judson Webb writes: " ... the more I tried to sort out and understand the arguments, to sift claim and counterclaim, the more I found that most of the central figures, however original they might have seemed, had really gotten key ideas from their teachers and predecessors...".[249] A historical observation can indeed help to reveal "the intentions behind the ideas"[250] we have just been concerned with.

As regards the central question Husserl investigates in his treatise "*On the concept of the operation*", namely 'which kinds of *equivalent* cognitions are to be explicitly expressed in formal sciences?'[251] the source of inspiration for the whole discussion may very well have been Leibniz's defence of the role of identities. Locke's mechanical spokesman[252] in the *Nouveaux Essais*, Philalèthe, calls them "*proposition frivoles*": he regards them as trifling because they can be seen "at first blush ... to contain no instruction".[253]

In Chapter 2 of book IV of his *New Essays* (*Des degrés de notre connaissance*) Leibniz tries to show the usefulness of identities, of primary truths of reason that "seem to do nothing but to repeat the same thing without telling us anything" (§1). Identities, according to Leibniz, can be either affirmative or negative. Under the first heading he includes

(i) Propositions expressed by instances of the schema "Every A is A" ("A $\subseteq$ A").[254] He calls them "completely identical propositions (*identiques qui le sont entièrement*)".[255]
(ii) Propositions exemplifying the schemata "Every A which is B is A" ("AB $\subseteq$ A") or "Every A which is B is B" ("AB $\subseteq$ B"). Leibniz calls them "semi-identical propositions (*identiques à demi*)".[256]
(iii) Propositions expressed by instances of schemata of hypothetical form such as "if every A is both B and C, then every A is B" ("A $\subseteq$ BC $\rightarrow$ A $\subseteq$ B").[257]
(iv) "Conjunction [and] disjunctions can likewise be identities." Here Leibniz seems to be thinking of propositions that instantiate the schemata "Everything which is A and (or) B is A and (or) B" ("AB $\subseteq$ AB"; "A+B $\subseteq$ B+A").[258]

---

[249] Webb 1980, xii.

[250] Loc. cit.

[251] See above §1.2 (question 2).

[252] I borrow here terminology from Remnant & Bennett 1996, 10.

[253] Leibniz 1704, book IV, ch. viii, §2.

[254] This formulation is taken from Künne 2009, 253.

[255] Leibniz 1704, book IV, ch. viii, §4. Kant calls them *tautologisch*, in Bolzano they are called *identisch oder tautologisch*. Cp. Künne 2009, 253, fn. 52 & 53.

[256] Leibniz 1704, book IV, ch. viii, §5.

[257] Leibniz 1704, book IV, ch. ii, §1.

[258] Loc. cit.

(v) Even instances of the schema "Every non-A is non-A" ("–A ⊆ –A") are subsumed under the heading 'affirmative identities'; for they also instantiate the negation-free schema "Every B is B".

Leibniz classifies as negative identities both *the principle of contradiction* and propositions he calls *disparities* (*disparates*). He formulates the former as the principle of bivalence: "Every proposition is either true or false", where the disjunction is taken to be exclusive. Disparities are propositions which say that "the object of one idea is not the object of another idea". Leibniz's examples are of the sort "Warmth is not the same thing as colour", "Man and animal are not the same although every animal is an animal". Presumably "Descartes is not Spinoza" would qualify, too.

After this preparation let us consider the question whether identities (trifling propositions) are useless for cognitive purposes. Leibniz writes:

> Someone who has been listening patiently so far to what I have just been saying will finally lose patience and say I am wasting time on trivial assertions and that identities are all useless. But this verdict results from not having thought enough about these matters. The inferences of logic, for example, are demonstrated by means of identities, and geometers need the principle of contradiction for their demonstrations by *reductio ad absurdum*. At this point let me merely show how identities can be used in demonstrating [the soundness of] some inferences of reason.[259]

He then goes on to prove both (i) that by means of the only principle of contradiction it is possible to obtain the second and the third figures of the syllogism from the first and (ii) that the conversions traditional logicians appeal to can be proved by means of the second and the third figures of the syllogism.[260]

Let us firstly consider (i). The inference "A & B, therefore C" is valid iff the inference "A & ¬C, therefore ¬B" is valid iff the inference "¬C & B, also ¬A" is valid. (In the derivation we take each time one of the premises to be true and the other premise and the conclusion to be false.) Two examples may suffice. From *Darii*, "MaP, SiM, therefore SiP" we obtain "MaP, ¬SiP, therefore ¬SiM". Then relying on the diagonals in the Square of Oppositions, we arrive at *Camestres*, "MaP, SeP, therefore SeM". Similarly, from *Ferio* ("MeP, SiM, therefore SoP") we obtain "¬SoP, SiM, therefore ¬MeP". Then, again relying on the diagonals in the Square of Oppositions, we arrive at *Datisi*, "SaP, SiM, therefore MiP".

Let us now consider (ii), i.e. the provability of the conversions of traditional logic, i.e. *conversio simplex*, "SeP → PeS" and "SiP → PiS", and *conversio per accidens*, "SaP → PiS" and "SeP → PoS", by means of the syllogisms of the second and of the third figure. (Note that the conversions *per accidens* depend on Aristotle's assumption that universal propositions have existential import.) Again two examples may suffice. In order to prove the simple conversion "SeP → PeS"

---

[259] Loc. cit.

[260] Leibniz acknowledges that the Parisian philosopher Pierre de la Ramée (Petrus Ramus) was already aware of (b).

we assume "SeP" and the *proposition frivole* "PaP" and infer "PeS" using the syllogism of the second figure *Cesare*, where "P" functions both as *terminus medius* and as *terminus minor* and "S" as *terminus major*. Similarly, in order to prove the conversion *per accidens* "SaP → PiS" we assume the *proposition frivole* "SaS" and "SaP" and infer "PiS" using the syllogism of the third figure *Darapti*, where "S" functions both as *terminus medius* and as *terminus major* and "P" as *terminus minor*. (This syllogism depends on Aristotle's assumption, for otherwise it would be possible to prove something that essentially depends on this assumption, namely a *conversio per accidens*, by something that does not depend on it.) At the end of this long discussion Leibniz stresses the importance of identities in formal sciences:

> This show that the purest identities, which appear entirely useless, are really of considerable use in abstract and general matters [i.e. in the formal sciences]; and that can teach us that no truth can be scorned

In book IV Chapter viii (*Des propositions frivoles*) Théophile repeats this point against Philalèthe whom he makes admit: "It seems that these identical maxims are merely trifling – or *nugatoriae*, as even the Scholastics call them. And I would not be satisfied with just saying that that *seems* to be so, had not your surprising example of the demonstration of conversion by interposition of identities made me step with care when it comes to being scornful of anything." Théophile first emphasizes the importance of the principle of contradiction (also a trifling proposition, as we saw) in apagogic proofs: "Do you count that as nothing, Sir? Do you not recognize that to reduce a proposition to absurdity is to demonstrate its contradictory?" He then renews the thesis he maintained in Chapter 2: "you can see quite well how identities should be used if they are to be useful – namely by showing that other truths which one wishes to establish can be reduced to them by means of deductions and definitions."

Let us now read Husserl's text against this Leibnizian background. He poses the question which kind of immediate equivalences are worthy of being expressed in abstract formal sciences? Are they useful for the purposes of reduction and/or derivation? And then his argument runs on Leibnizian lines:

> [W]e do not need to engage in the controversy over whether *immediate equivalences* ... considered in and for themselves must be regarded as extensions of knowledge or not. In any case they *can* serve as extremely important instruments for unquestionably extending our knowledge, namely, for leading it beyond the domain of immediate equivalents. This will prove true in all cases where a proposition - through immediate equivalent transformation – first receives that form which makes it appear as an appropriate premise for certain inferences. The possibility of an inference can impose itself if the premises are of a certain form... The equivalent transformations that show up under the "immediate inferences" of traditional logic can serve as examples.[261]

[261]*PdA* App. 412–413; *PoA* 389–390 (my emphasis).

1.18.2 On the Notion of Computation and on Boole

At the end of his complex discussion of the "calculational technique (*Rechenkunst*)" as opposed to the science of arithmetic (*Arithmetik*), Husserl arrives in Chapter XII of his book at a characterization of the notion of *computation* (*Rechnen*) or *calculation* that he finally deems satisfactory: a computation is "any rule-governed mode of derivation of signs from signs within any algorithmic sign-system according to the 'laws' – or better: the conventions – for combination, separation and transformation peculiar to that system."[262]

What is really interesting in this characterization lies mainly in the explicit acknowledgement of the *generality of the concept* under scrutiny, more precisely, in the fact that the notion of computation is completely uncoupled from that of *number*, and more generally from that of *quantity*, in which the whole issue had originated. In Husserl's words, "there are higher logical interests than those of *arithmetica numerosa* (with which we currently have to do) which require this delimitation of the concept."[263]

Thanks to the generality of this characterization it is possible to analyze and represent the structure of *any* 'problem-solving' process – and not just of those of numerical kind, but independently from the specific nature of the domain in which we are operating – in three distinct and sequential moments, the first and third of which are *conceptual*, while the second is *purely formal-algorithmic*:

formal encoding of the problem
("conversion of the initial thought into signs")
↓
calculation
↓
solution as decoding of the result of the calculation
("conversion of the resulting signs back into thoughts")

A careful reader cannot fail to notice the similarity (with just one important difference, to which we will return below) of this abstract representation of the problem-solving processes with the analysis of "symbolic reasoning" proposed by George Boole in the *Laws of Thought*.[264] Diagrammatically (as can be seen from Chapter V, esp. §4), according to Boole a (correct) *symbolic reasoning* can be represented by the composition of three moments or steps:

---

[262] *PoA* 273; *PdA* 258.

[263] Loc. cit.

[264] Boole 1854.

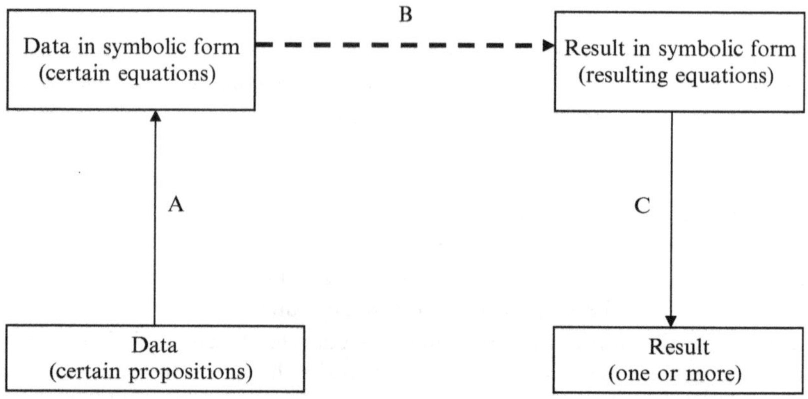

- The first moment (A) consists in the encoding of the data in symbols: these must have a well-defined *interpretation*, and their laws of combination must be determined by that interpretation.
- The second moment (B) consists in the application of *symbolic-formal processes* conforming to the laws of combination, and it is completely independent from the requirements of interpretability: Boole explicitly allows the *possibility of non-interpretability of the intermediate steps in calculation.*
- The third moment (C) finally consists in the *interpretation* (decoding) of the result of the symbolic process on the basis of the 'coding system' chosen for the symbolization of the data.

Admittedly, in Chapter XIII of the *Philosophy of Arithmetic* Boole is not mentioned (indeed, he is not mentioned at all in the entire work). However, in Husserl's lecture "*Über die neueren Forschungen zur deduktiven Logik*"[265] of 1895, we find an entire section (circa twenty pages) devoted to Boole. The main part of this section (which is the one we want to outline briefly here) is focused on the notion of *calculus.*

Husserl's considerations begin with the observation that the real strength of Boole's work does not reside in his proposed logical analysis of language (on the contrary: "he has been less concerned about a possibly complete analysis of the forms of judgment"[266]), but rather in the development of a *logical calculus, i.e. the development of logic as a computational discipline*:

> He took ... and accepted from his predecessors, from Hamilton and in part also from De Morgan, what he could use. Use for what? The answer is: for the development of calculus (*für die kalkulatorische Entwicklung*). For this is the goal that he had set for himself from the beginning and that he pursued with powerful genius and achieved with full certainty: establishing formal logic as a mathematical science. As arithmetic is the computational

[265]*LV 96*, App. 305–328.

[266]*LV 96*, App. 305.

discipline about numbers, so formal logic is to be developed as the computational discipline about concepts and states of affairs (*kalkulatorische Disziplin von den Begriffen und Sachverhalten*).[267]

At this point, Husserl puts Boole aside (returning to him only in the final pages, where he summarily describes Boole's class calculus) and develops an articulated and very interesting discussion of the concept of *calculus*, using arithmetic as a paradigm.

Of the four elementary arithmetical operations on natural numbers (greater than zero), addition and multiplication are total, i.e. defined for every natural number, while subtraction and division are partial ("subtraction: a – b. Defined, if a > b; and then also univocal").[268] Consider the following nine fundamental laws (*Fundamentale Gesetze* or *Grundgesetze*):

1. $a + b = b + a$
2. $(a + b) + c = a + (b + c)$
3. $(a - b) + b = a$
4. *if* $a + c = b + c$ *then* $a = b$
5. $ab = ba$
6. $a(bc) = (ab)c$
7. $a(b + c) = ab + ac$
8. $(a/b)b = a$
9. *if* $cb = db$ *then* $c = d$

The concept of addition, and mediately the *concepts* of subtraction, multiplication, division, *are founded* on the *concept* of number; and the nine laws listed above *derive a priori from these concepts*. All remaining true propositions of arithmetic that exclusively concern these four operations of the calculus are *purely formal* deductions from the principles (1)–(9). More exactly, they can *be proved without appeal to the concepts* (of number, of the four operations), by manipulating the "sensuous expressions on paper" corresponding to them in accordance with the *rules of symbolic transformation* corresponding to the principles (1)–(9). For example, principle (1) corresponds to the rule of transformation that allows the equivalent substitution of every sign construction of the form 'a + b' with the construction 'b + a'.

> To prove any further proposition I do not need at all to fall back on the concept of number or on the concepts of addition, multiplication, etc. The proof rather consists entirely of steps in which we do not need to do anything but apply one of the nine propositions, i.e. we have to do nothing but subsume. The nine basic propositions (*Grundsätze*), however, are formally independent from each other... In order to grasp (*einsehen*) them we must appeal again to the concepts of number or addition etc. We cannot represent any of these propositions as a special case of the others or derive it from them through transformations.[269]

---

[267] *LV 96*, App. 306.

[268] *LV 96*, App. 307.

[269] *LV 96*, App. 308.

Calculating is a *non-conceptual* process that operates on complexes of signs and proceeds formally according to pre-established rules of transformation. And it is clear, Husserl claims, that there is no *a priori* reason to limit calculating to the numerical domain: in general it is possible to apply the calculus, an algorithmic system, to any domain that allows an 'algebraic' structuring:

> It is immediately clear that there is no a priori reason why calculating should be restricted to the arithmetical domain. Wherever we find a domain of concepts in which obtain analogous relations as in that of arithmetic, i.e. wherever we can find uniform ways of constructing new concepts from given ones, in such a way that the results of the constructions can always serve as elements for new constructions, and where there is a limited number of laws for these kinds of construction, an infinite manifold of pure theorems (*Folgesätze*) is deducible from the axioms (*Grundsätze*), and this in the way of a purely formal deduction. And the computational way of proceeding will also be possible that makes falling back on the concepts superfluous and relies only on the external forms of the process.[270]

Here we have in a nutshell the essence of calculation:

> What is characteristic of calculation? Nothing but the fact that it is a procedure for deducing equivalent propositions from certain given propositions in a research domain, without falling back on the specific concepts and relations. How is this possible? What are we dealing with, if not concepts and relations? The answer is: the concepts are thought by means of certain terms, the relations by means of corresponding connecting signs. When we calculate in arithmetic, we only care about the signs and the rules of their connection.[271]

This characterization is perfectly in line with what emerged at the end of the discussion on computation in Chapter XIII of the *Philosophy of Arithmetic*. But in Husserl's reflection on Boole we also find further interesting remarks, two of which are worth considering in the present context.

Husserl underscores repeatedly, and with abundant examples, the importance of generality and flexibility of systems of calculation for their application: an algorithmic system obtained – as the one considered at the beginning – by *abstraction* from a specific conceptual domain will also be interpretable in *different* though structurally similar conceptual domains. This results in obvious "savings in deductive labour": a formal proposition, once deduced by calculation, translates into a true conceptual proposition in every domain that falls under the algorithm. But the most significant aspect here is that Husserl explicitly takes into account also the process opposite to one exemplified above (schematically: determinate conceptual domain → abstraction → algorithmic system), i.e. the process (which is the essence of the "axiomatic revolution" of the late nineteenth century) that consists in *starting* from the constitution of an algorithmic system, and *then to* look for possible *interpretations* (possible *models*).

> But also another way is open. One develops the algorithm for itself and says: every conceptual domain which is such that we can designate its basic concepts (*Grundbegriffe*)

---

[270] *LV 96*, App. 312.

[271] *LV 96*, App. 309.

> by the basic signs, its concepts of combination (*Verbindung*) by the signs for combinations of the algorithm – obviously to the basic rules (*Grundregeln*) in the domain there correspond basic laws (*Grundgesetze*) in the algorithm – falls under the algorithm with respect to all of its deductions.[272]

While in the case of algorithmic systems obtained by abstraction we have to do with *synthetic a priori* laws, founded *in given concepts*, in this second case the laws are *purely analytic*, and the concept is obtained through reflection on the form of the law.

Let us proceed to the second point that we deem relevant here. Suppose we set up an algorithm (i.e. a system of signs plus rules of transformation): do the signs have a meaning which is not the original, conceptual one, if the algorithm is constituted by abstraction from a specific conceptual domain)? And if so, what is it? Husserl's answer to this question is affirmative: the signs do (still) have a meaning in the algorithm, namely an *operational* meaning, a "*Spielbedeutung*", which is determined by the complex of formal rules that govern its manipulation.

> Suppose a kind of given signs is set and memorized and so is a certain number of rules that like the rules of a game determine how we are allowed to operate with the signs, in such a way that every other way of proceeding is considered unacceptable. Then an arbitrary connection of signs can, on the basis of the rules, be replaced by various equivalent connections of signs. And a derivation is correct if all of its steps are in accordance with the rules, i.e. no step is taken which does not have a justification by a simple subsumption under one of the rules. Hence if I consider the signs in this way by themselves, they are not merely doodles on the paper, they clearly have a certain meaning. What, then, is their meaning? It is no longer the corresponding arithmetical meaning, because I have wholly abstracted from it. Clearly the meaning now lies in the rules of the game. It is exactly like as in the game of chess: the bishop, castle etc. Now I maintain: all calculating consists in the fact that the original concepts, the concepts of number and the concepts of relation and connection belonging to them, are replaced by their mere symbols and these are now considered only as such purely conventional game-concepts. The game-meaning (*Spielbedeutung*) of these symbols then lies in certain rules which are nothing but the exact counterparts of the fundamental laws to which all arithmetical deduction can be reduced by mere subsumption.[273]

There is a stark contrast here with Frege's attitude in his paper *Ueber formale Theorien der Arithmetik* (1885).[274] In the course of his criticism of Heine's and Thomae's theory of irrational numbers in the second volume of his *Grundgesetze* (1903) Frege concedes that the *Begriffschrift*, too, can be conceived as a *Spiel* (§90) but he doubts that a theory of *Rechenspiele* is possible (§93).

Husserl appends a brief description of the Boolean calculus to his general reflections on the concept of calculus. This calculus is presented as a calculus for the domain of classes in general: "Boole originally constructs his calculus as a

---

[272] *LV 96*, App. 314.

[273] *LV 96*, App. 310.

[274] Frege 1990, 103–111.

calculus of classes. As arithmetic is the calculus for the field of numbers (*Anzahlgebiet*), so the Boolean calculus is the calculus for the field of classes in general (*im Allgemeinen*)." While this expository part is not really worth discussing a remark should be made on Husserl's (and not only on Husserl's) fundamental critique of Boole, as regards the *correctness* of the Boolean method. One can summarize the entire discussion as follows. Boole explicitly admits (and freely avails himself of) the possibility, at the risk of losing the generality of the calculus, that not all the steps in the symbolic-formal process that leads from the encoded data to the to-be-decoded result are *interpretable*; in other words, he does not require that there be a parallelism between the conceptual and the symbolic-algorithmic level *along the entire symbolic process*: the important factor is that this parallelism exists at the beginning (*input*) and at end (*output*). Husserl's position appears to be the opposite, i.e. that the parallelism between the two levels should be *constant*: "Finally, to every principle corresponds a certain rule for operating with the signs, and every derived proposition is obtained by mere stepwise subsumption under these rules. Thus there is a one-to-one corresponding parallelism between the game-system and its rules and the number-system and its laws. Hence there is no mechanically-symbolically derivable proposition that does not have its counterpart in the domain of arithmetic."[275] Similar considerations are to be found in "*On the concept of the operation*", in the context of Husserl's discussion of algorithmic systems:

> Whoever has ... has attained clarity about the algorithmic methods that run precisely parallel to the ... objective [*sachlichen*] methods (operating on the concepts themselves ...) will see that all immediate equivalences that mediate (*vermitteln*) in the objective methods must have their counterpart in formulae that mediate in the algorithmic methods.[276]

Husserl's conclusion is that in Boole the strength of the idea of extending the notion of calculation beyond the sphere of quantity is coupled with an intrinsic weakness at the level of the foundations of the computational process.

> Hence Boole's method must have appeared like shadow boxing (*Spiegelfechterei*, i.e. a sham), where, however, one had to register the unexplainable miracle that Boole's calculations always led to correct results: somebody who trusts the Boolean method and solves a logical task by calculatorial means, would in fact find a true solution, while one would have expected that a meaningless method would also deliver meaningless or at least false results. However, Boole himself did not have an entirely clear conception of the reasons for the validity of his method. In his case we are dealing with a brilliant intuition rather than with a conceptual insight. The logical principles of the calculatorial method remained completely precluded to him as well as to later researchers.[277]

---

[275] *LV 96*, App. 311.

[276] *PdA* App. 414; *PoA* 391.

[277] *LV 96*, App. 322–323.

1.19 Appendix 3: Sets and Finite Numbers in "Zur Lehre vom Inbegriff"

1.19.1 Introduction

According to the editor, the treatise "*Zur Lehre vom Inbegriff* " contains mostly preparatory research notes for the second volume of the *Philosophy of Arithmetic*, that were written around the end of 1891.[278] This cannot be right. The text contains several indications that imply it is *later* than 1891, by at least a few years. First, there is an *explicit* reference to a result obtained by Felix Bernstein[279] who was only 13 years old in 1891.[280] Further evidence can be gleaned from the discussion of a specific problem, i.e. the issue of "comparability" of cardinal numbers, which we will discuss in detail below. Finally, there is a note (written in pencil, which could have been added later) that contains a reference to a publication by E. Schröder of 1898.[281]

"*Zur Lehre vom Inbegriff*" is not Husserl's title (on the external cover of the manuscript there is the title 'Formal arithmetic', written in pencil), but has been added by the editor of the *Husserliana* edition based on the content of the manuscript. Indeed, at first sight the study presents itself as Husserl's attempt to elaborate a *set theory* in Cantor's sense. From an annotation in the margin of the manuscript[282] we know that Cantor read the work; moreover, given the very friendly personal and scientific relationship among them, we can suppose that Husserl might have received an impulse from Cantor in the direction of the elaboration of these reflections.[283]

We have to stress right from the start that the explication of the general notion of set to which the initial pages of the study are devoted, actually serves to pave the way for the discussion of the real topic of the investigation: "sets of units," i.e. *numbers*. Husserl's real intent is clearly to obtain not only an adequate definition of

---

[278] *PdA* App. 385–407; *PoA* 359–383; *Textkritische Anmerkungen*, 530–533.

[279] "Bernstein has demonstrated ... a sufficient condition ... On this point we still must have an exchange with Bernstein." *PdA* App. 394; *PoA* 369.

[280] Felix Bernstein (Halle 1878 – Zürich 1956) studied with Cantor in Halle, then with Hilbert and Klein at Göttingen, where he obtained his doctorate with a dissertation entitled *Untersuchungen aus der Mengenlehre*. After his habilitation (Halle 1903), Bernstein taught at Göttingen from 1907 to 1932. After moving to the United States he taught for sixteen years at various universities (Columbia University, New York University, Syracuse University), and in 1949 returned to Göttingen. According to various biographic notes, Bernstein already started to follow Cantor's seminars at the university of Halle while still attending the gymnasium; on the other hand, a precise indication of the year in which contact with Cantor began is unavailable (this information is not to be found in the most complete biography: Frewer 1981). However, it is reasonable to suppose that it did not begin before 1894–95 (Bernstein obtained his *Abitur* in 1896).

[281] "Read to *Cantor* when he told me of a treatise of Schröder's for the *Leopoldina*" (*PdA* App. 399, n. 1; *PoA* 374, n. 9). The reference is to Schröder 1898.

[282] Loc. cit.

[283] Cp. Ortiz Hill 1994a, 3 & b, 96 & 1997b, 137 ff. & 2004, 110–114.

the general concept of number, but also a rigorous and systematic foundation for the theory of *finite* cardinal numbers, so as to correct two – closely connected – fundamental "omissions" that he finds in his *Philosophy of Arithmetic*:

(i) A precise determination of the dichotomy *finite*/*infinite* with respect to sets, and hence also with respect to cardinal numbers
(ii) An acknowledgement of *equivalence* (i.e. of the relation of 'standing in one-to-one correspondence') as an indispensable criterion for ordering improper number presentations according to the order relation

The central stages of the study, which are the most complex and difficult ones, are devoted to the detailed presentation (also from a "technical" point of view) of a rigorous foundation of the theory of finite numbers. In the concluding pages Husserl corrects some of the positions of the *Philosophy of Arithmetic*, specifically the rejection of the definition of number through equivalence. Put in a nut-shell, Husserl's argumentation is this: If we limit ourselves to the consideration of *proper* number presentations, a definition of equality through equivalence is not necessary, because we can distinguish proper presentations of numbers and immediately classify them according to the order relation, and hence it is not necessary to compare numbers according to equivalence. But for *improper* number presentations equivalence is indispensable for a classification based on order. How can we distinguish two symbolic presentations of number? How can we assess whether two numbers, presented symbolically, are equal? We have two options:

(1) We first define the natural number series and then the concept of finite number as set of units that can be associated by bijection to an *initial segment* of the series of naturals (though we have to point out that the one-to-one correspondence is independent from the order given to the set of units)
(2) We define numerical equality *via* the relation of equivalence, and then the notion of finiteness using that of equivalence (like Dedekind)

Whichever of these roads one might take (Husserl developed only the second one, as we shall see), the role of the notion of one-to-one correlation, i.e. bijection, is clearly essential.

Here it may be useful to recall that the proof of the *equivalence* of the two notions of *finiteness* involved in (1) and (2) above (finiteness of a set X as *the possibility of bijection of X with a natural number* – $\mathrm{FIN}_1(X)$ – and finiteness of a set X as *impossibility of bijection of X with any of its proper parts* – $\mathrm{FIN}_2(X)$ or "Dedekind-finiteness") requires the *axiom of choice*.[284] More precisely, this axiom is necessary to prove that from $\mathrm{FIN}_2(X)$ follows $\mathrm{FIN}_1(X)$, while it is not necessary for the converse implication, the so-called "fundamental theorem of finite arithmetic": $\mathrm{FIN}_1(X) \Rightarrow \mathrm{FIN}_2(X)$.[285]

[284]In one of the many possible equivalent formulations: *For every family F of non-empty and disjoint sets, there is at least one set X having one and only one element in common with each of the sets belonging to F.*

[285]See Tarski 1925.

1.19.2 Sets and Operations on Sets

In "*Zur Lehre vom Inbegriff* " the *sets*, or *aggregates*, are objects designated by terms of the form 'A and B and C ...' that can be given either as objectualizations of an act of thought or intuition, proper or symbolic, or as extensions of a property. The aggregate, as mathematical object, however, is independent from the way it is given as well as from the way in which its single elements A, B, C, ... are given. "If 'A', 'B', 'C', ... designate any objects whatever, whether intuited or thought, existent or imaginary, ... then the expression 'A and B and C and ...,' taken in its general sense, yields a definition of the term 'collection (*Inbegriff*)'."[286]

As in the *Philosophy of Arithmetic*, Husserl maintains that the concept of a collection has its psychological origin in a collecting act that "binds" different objects in an "ideal unity," but he observes that "conceptual determinations can be given that decide in a general manner which objects are and which are not to belong to the intended unity ... for example, when we speak of a collection of objects that fall under a concept C."[287] As to the members of the collection, "for logical purposes ... it does not matter whether the objects which are to be grasped together are intuited separately, along with their individual peculiarities, or are only represented (*repräsentiert*),"[288] i.e. it is sufficient to have an improper presentation of the collection.

In this respect, it appears possible to maintain that Cantor's distinction between *internal determination* and *external determination* of a manifold (see his paper 1883) is at least an *aspect* of the distinction that we find in Husserl between properly and symbolically presented aggregates. Of course, a properly presented collection is a manifold of which we can distinctly perceive each member while an improperly presented collection is a manifold for which this is not possible. In "*Zur Lehre vom Inbegriff*" Husserl declares explicitly that improperly presented collections are sufficient for the aims of logic. For the sake of simplicity, let us consider only the second way in which a collection can be given, i.e. as the extension of a concept. On the basis of the definition of the concept it remains determined for each object whether it belongs to the collection or not, independent of the existence of an effective method to *establish* whether or not the object belongs to the collection. Now, for Cantor, a manifold is well-defined if the preceding condition is met and if, furthermore, given two objects belonging to the manifold, it is possible to decide on the basis of the definition whether they are the same or not. Cantor explicitly states that "in general *relative decisions will not be effectively executable* exactly and with certainty on the basis of the available methods and capacities – but this is not relevant; only the internal determination is important; this can then be transformed into an external determination ...," which means: transformed into an effective procedure to decide if the object belongs to the manifold or not. As an example Cantor gives the definition of the set of all algebraic numbers. This constitutes the

[286] *PdA* App. 385; *PoA* 359.

[287] *PdA* App. 385 f.; *PoA* 359 f.

[288] *PdA* App. 385; *PoA* 359.

internal determination on the basis of which any number r belongs or not to the class of algebraic numbers; however, the problem of actually producing a decision for a given number r often turns out to be very difficult. In other words, in the classical conception a set M is considered given when for each object it is univocally determined whether or not it belongs to the set. The possibility of effectively solving every problem of the form "Does x belong to M?" is a different problem.

Bolzano, in his *Paradoxien des Unendlichen* (1851), was apparently the first to introduce the concept of a set in an *extensional* sense: "I call a set (*Menge*) a collection (*Inbegriff*) which we put under a concept so that the arrangement of its parts is unimportant (in which therefore nothing that is essential for us changes if only this arrangement changes)" (§4).[289] Sets, on this view, are objects exclusively characterized by their elements: they are said to be equal if they have exactly the same elements (*principle of extensionality*).

The definition of the notion of set given by Husserl is very close to Cantor's and to Dedekind's notion of 'system'.[290] Like Dedekind and Cantor Husserl proposes a "naïve" approach to the concept of set: on the one hand, there is an appeal to the faculty of the mind of "uniting," of "thinking together" or of "correlating things with things," and sets are conceived as objectualizations of the creative acts of thought; on the other, the problem of identifying the explicit logical principles that govern the manipulation of these concepts is avoided. In this respect, it is important to recall that in more or less the same period Frege undertook a project that aimed at systematically reducing the concepts of set and number to pure logic.

After describing the notion of set in the terms outlined above, Husserl moves on to a list of *axioms* concerning sets and certain set-theoretical operations that he introduces in that context. In order to improve occasionally on his exposition we will use the following symbols:

- $\in$, for the membership relation between an object and a set;
- $\subseteq$ ($\subset$), for the relation of inclusion (resp. *proper* inclusion) between sets;
- V, for the total set;
- $\equiv$, for identity, i.e. extensional equality, among sets.

It is important two keep things in mind:

1. Husserl excludes the possibility of *sets with less than two elements*.[291] Hence, there is no empty set, and there are no singletons, i.e. sets of the form $\{x\}$, for a certain object x. However, in order to formulate the axioms about the operations

[289]Bolzano 1975. Cp. *WL* I, §§84–86.

[290]"By a 'set (*Menge*)' we are to understand any collection into a whole (*Zusammenfassung zu einem Ganzen*) M of definite and separate objects m of our intuition or our thought. These objects are called the 'elements' of M" (Cantor 1895, 481; transl. 1955, 85). "It very frequently happens that different things $a, b, c, \ldots$ for some reason can be considered from a common point of view, can be associated in the mind, and we say that they form a *system* S" (Dedekind 1888; 1–2).

[291]This condition on the concept of collection is taken from § 82 of Bolzano's *Wissenschaftslehre*.

on sets without using too many distinctions, he establishes the convention of *treating individuals as if they were sets with just one element.*[292] In other words, he conventionally identifies x with $\{x\}$ (and consequently $x \in$ A with$\{x\} \subseteq$ A), and he uses capital letters indifferently to denote sets and single objects. We prefer to use the majuscules 'A', 'B', ... to refer to sets (including singletons), and the minuscules x, y, ... to refer to objects.

2. When Husserl says that a set A is 'part' (*Teilinbegriff*) of a set B, he normally means a *proper part*. Hence:

   - $A \subseteq B := \forall x(x \in A \rightarrow x \in B)$ (inclusion);
   - $A \equiv B := A \subseteq B \wedge B \subseteq A$ (*extensional* equality);
   - $A \subset B := A \subseteq B \wedge \neg(B \subseteq A)$ (*proper* inclusion).

Husserl individuates four operations on sets: 'augmenting (*Vermehrung*)', 'diminishing (*Verminderung*)', 'connection (*Verknüpfung*)', and 'partition (*Teilung*)'. We note that the first two are not operations in the sense of *univocal correspondences*:

- *Augmenting* a collection is uniting its objects with *one or more new objects* and to constitute a new collection with these;
- *Diminishing* a collection is removing *some* of its objects and uniting the remaining into a new collection.

The two remaining operations correspond to the *addition* and *subtraction* of sets:

- *Adding* a collection B to a collection A, given that A and B are *disjoint* (i.e. they have no common elements) is *augmenting* A with the objects of B. The result of this operation is designated by 'A + B'. More generally, given certain sets that are pairwise disjoint A, B, C, ..., 'A + B + C + ...' designates the comprehensive set that contains the objects of A, of B, of C ... taken together.
- *Subtracting* a collection B from a collection A, given that B is a (*proper*) part of A, is *diminishing* A with the objects of B. The result of this operation is designated by 'A − B'.

So, addition and subtraction are taken (*à la* Boole) as *non-total* operations. The "*axioms*" (understood as evident truths) are now the following six:

[HU.1] "For every collection with the single exception of that one which includes everything representable in the widest sense of the word, there is a possible further object which is not contained in it."[293] In symbols:

$$\neg(A \equiv V) \rightarrow \exists x(x \notin A).$$

[292] *PdA* App. 387; *PoA* 361.

[293] *PdA* App. 386 ff.; *PoA* 360 ff.

Though Husserl says that he will make essential use of this axiom later on, we note that it is nothing but a kind of "definition" of the total set V. Indeed, formally speaking, HU.1 is equivalent to saying that if the set A contains all objects, then A coincides with the total set:

$$[\neg(A \equiv V) \rightarrow \exists x(x \notin A)] \leftrightarrow [\neg\exists x \neg(x \in A) \rightarrow A \equiv V] \leftrightarrow [\forall x(x \in A) \rightarrow A \equiv V)].$$

[HU.2] "It is evident that every totality can be augmented by an arbitrarily selected object not contained in it." In symbols (using operation +):

$$\forall x(x \notin A \rightarrow \exists B(B \equiv A + \{x\}))$$

It follows from [HU.1] and [HU.2] that every set A such that $\neg(A \equiv V)$ can be *augmented.*

[HU.3] "To augment (or diminish) a totality by certain objects, and to diminish (or augment) the resulting totality by identically the same objects restores the original totality. In other words, augmentation and diminishment are inverse operations." In symbols (using the operations of addition and subtraction):

$$\forall x(x \in A \rightarrow x \notin B) \rightarrow (A + B) - B \equiv A,$$
$$B \subset A \rightarrow (A - B) + B \equiv A.$$ [294]

Hence, addition and subtraction of sets are the inverse operations of each other.

[HU.4] $A + B \equiv B + A$.

[HU.5] $(A + B) + C \equiv A + (B + C)$.

These two axioms establish that addition is *commutative* and *associative* (when it is definite).

[HU.6] "Any totality admits of being diminished by one unit." In symbols:

$$\forall x(x \in A \rightarrow \exists B(B = A - \{x\})).$$

Formulated this way, the axiom is valid, obviously, because of the hypothesis that every set has at least two elements and because of the conventional identification of x with $\{x\}$.

[294]Husserl rewrites axiom [HU.3] so: $(I + I') - I' \equiv I$, $(I - I') + I' \equiv I$. But he observes that if I is a collection and I′ is either a single object or a collection, then "I + I′" is meaningful if and only if I′ does not belong to I, resp. if and only if I and I′ are disjoint. Analogously, "I – I′" is meaningful only when I′ is a proper part of I.

1.19.3 Definition of the General Concept of Cardinal Number

A *cardinal number* is here defined by Husserl as a 'collection of units'. The number is obtained starting from a collection of objects, completely abstracting from every concrete determination of the content of such objects, and taking into account only the fact that among themselves they are, in some way, distinct. This abstraction is possible for every collection in general; in other words (and using Husserl's notation), every collection $I_A(A, B, C, \ldots)$ of arbitrary objects is always correlated with a collection *of units* $I_1(1, 1, 1, \ldots)$ obtained from the first by replacing each element with "something" (*Etwas*) or "one" (*Eins*), here indicated with '1'. The number that belongs to a collection I of arbitrary objects is that particular collection that is obtained from I by *abstraction and transition to units*.

It is noteworthy that Husserl also considers a variant of the process of abstraction, described above, that generates the cardinal numbers. This new abstraction does not lead to the pure *Anzahlen* but to the so-called '*benannte Zahlen*', a term that one might translate as 'qualified numbers'. Given any group of objects that are all *of the same kind*, B, we abstract from all the characteristics of the elements, except for the fact that they are distinct and are of kind B. In this way we do not obtain a set of units, but a set of *units of kind* B, called '*benannte Einheiten*'. The relation between *reine Anzahl* and *benannte Zahl*[295] can be thought of in two ways: one can consider a *benannte Zahl* as a specification of the concept of *Anzahl*, obtained by adding to the units of pure *Anzahlen* the qualification of being units of kind B. One can also, inversely, consider pure *Anzahlen* as abstraction from *benannte Zahlen*, obtained by taking away the sortal term from the latter.

In any case, the way in which the concept of cardinal number is obtained constitutes, again, a specific element of connection with Dedekind as well as with Cantor. In *Was sind und was sollen die Zahlen?*,[296] Dedekind defines natural numbers as the elements of a simply infinite system in which the concrete character of the elements is neglected and only their discernibility is preserved,[297] while Cantor defines the "power or cardinal number" of a set M as "the general concept which, by means of our active faculty of thought, arises from the collection M when we make abstraction of the nature of its various elements *m* and of the order in which they are given."[298] Cantor, furthermore, observes explicitly, using a terminology nearly identical to Husserl's, that if one abstracts from the specific features

[295] Husserl alludes at this point (*PdA*. App. 389 n.) to Bolzano. Cp. now Bolzano, BGA 2A, 8, 15ff.

[296] Dedekind 1888.

[297] "With respect to this process through which we free the elements from every other content (abstraction), we can correctly affirm that numbers are a free creation of the human mind" (Dedekind, op. cit.).

[298] Cantor 1895, 481; transl. 1955, 86.

of each element of the set one obtains a *unit*, thus the cardinal number is a set of units that "has existence in our mind as an intellectual image or projection of the given collection M".

1.19.4 Comparison of Two Sets Relative to Their Cardinal Number

Two sets A and B are said to be "equal relative to their cardinal number (*in Hinsicht auf die Anzahl gleich*)" if and only if B can be obtained from A by transforming the members of the latter into members of the former in such a way that the following two conditions are met:

(i) Every member must remain a 'one' (*eine Eins*)
(ii) Different members must be always transformed into different members

Equivalently: there has to be an arbitrary law f that puts A in bijection with B, i.e.:

1. f is a function from A to B
2. f is injective: $\forall xy \in A(f(x) = f(y) \rightarrow x = y)$
3. f is surjective: $\forall y \in B\ \exists x \in A(f(x) = y)$

Given two sets (and hence also two *numbers* or sets *of units*) A and B, and using (as Husserl does)

- '$A \cong B$' to say that A and B are *equivalent*, i.e. that there exists a bijection between A and B,
- '$A = B$' to say that A and B are equinumerous, or equal with respect to their number,

we have

$$A = B \text{ if and only if } A \cong B.$$

Like Cantor, Husserl concludes that two sets have the same cardinal number if and only if they are *equivalent*: equivalence between sets constitutes the necessary and sufficient condition for the equality of their cardinal numbers. However, Husserl explicitly states that it is not the case that equivalence determines the equality of two cardinal numbers, thereby confirming at least in part the position of the *Philosophy of Arithmetic*. Equivalence is a "criterion" whose meaning consists in the fact that it constitutes "an irreplaceable means to classify the numerical field" in its totality.

Defined in this way, the cardinal number of a set turns out to be an *invariant* of the set with respect to the system of all the possible *permutations* of its elements: every permutation of the elements of the set is equivalent to every other permutation of the same. At this point Husserl makes an interesting reference to the fact that Bernstein has proven that a sufficient but not necessary condition for the validity of this second characterization of the concept of number is that "the group (*Menge*) of the permutations of the given group 'exists'. There are, we should mention, groups where the group of the permutations contains a contradiction."[299]

As to the relationship between the two relations of *extensional identity* ($\equiv$) and *numerical equality* between sets (=), the following (easily verifiable) theorem is put forward:

1. $A = B \wedge B \equiv C \rightarrow A = C$
2. $A \equiv B \wedge B = C \rightarrow A = C$

Again, in terms that are very close to Cantor's, Husserl maintains the necessity, given two sets with different cardinal numbers, to *prove* that those cardinal numbers are in a well determined relation of order (i.e. that one is *greater* than the other or the other way around).

If A and B are two sets that have a different cardinal number, i.e. if it obtains that $A \neq B$, *it is not at all obvious* – as Husserl underscores – that one of the following two situations is the case:

– Either A is equal to a part of B ($A \leq B$, in symbols), i.e. there exists a $C \subseteq B$ such that $A \cong C$, or
– B is equal to a part of A ($B \leq A$).

In effect, the implication under consideration:

$$A \neq B \rightarrow A \leq B \vee B \leq A$$

is nothing but a logically equivalent reformulation of the so-called *property of trichotomy*:

$$A = B \vee A \leq B \vee B \leq A;$$

i.e., still equivalently,

$$\textit{at least one among } A = B, A < B, B < A \text{ obtains,}$$

[299]*PdA* App. 394; *PoA* 369. Interpretation: supposing that the class of all the permutations would always be a set leads to a contradiction. In other words, Husserl shows here that he is aware of the need to distinguish between proper multiplicities (sets) and multiplicities that are "inconsistent" or "too extended" to be considered objects. Keep in mind that Cantor, already in 1895, identifies the paradox that Burali-Forti will make known in 1897, writing about it to Hilbert (1896) and Dedekind. In particular, in the famous letter to Dedekind of 1899, Cantor indicates the origin of certain difficulties that had been found in set-theory (specifically, the Burali-Forti paradox) due to the missing distinction between "absolutely infinite" or "inconsistent" multiplicities, for which "the assumption that all of its elements 'are together' leads to a contradiction, so that it is impossible to conceive of the multiplicity as a unity, as 'one finished thing'," and consistent multiplicities or sets for which the totality of the elements "can be thought without contradiction as 'being together', so that their collection into 'one thing' is possible" (Letter to Dedekind, 28 July 1899, in Cantor 1932, 443; transl. in van Heijenoort 1967, 114).

where $A < B =_{df} (A \leq B) \wedge (A \neq B)$. It is known that to prove the property of trichotomy, i.e. that any two sets A and B are always *comparable with respect to their number*, it is necessary to use the *axiom of choice*.[300]

Moreover, Husserl points out that the concepts of 'greater' and 'smaller' imply that they exclude each other: if $M < N$ then $\neg(N < M)$. This is equivalent, although the proof of this fact is *not at all straightforward*, to the validity of the property of *antisymmetry* for the relation $\leq$:

$$M \leq N \wedge N \leq M \rightarrow M \cong N,$$

i.e. again equivalently, the fact that

$$\textit{at most one among } A = B, A < B, B < A \text{ obtains.}$$

This is in effect what is affirmed by the so-called *Theorem of Cantor-Schröder-Bernstein*, whose proof is in turn anything but trivial.[301]

Compare the analogous specification by Cantor of these two properties[302] (in reverse order):

> We have seen that of the three relations $a = $ b, $a < b$, $b < a$ each one excludes the two others. On the other hand, the theorem that, with any two cardinal numbers a and b, one of those three relations must necessarily be realized, is by no means self-evident and can hardly be proved at this stage. Not until later, when we shall have gained a survey over the ascending sequence of the transfinite cardinal numbers and an insight into their connexion, will result the truth of the theorem: <A.> if a and b are any two cardinal numbers, then either $a = b$ or $a < b$ or $a > b$.[303]

---

[300]Cantor *asserts* the property of trichotomy, *without proving it*, already in his 1878. In the first of his 1895 papers he explicitly acknowledges the need for (and the difficulties of) a proof of this property (see next footnote). In the letter to Dedekind of 28 July 1899, Cantor sketches a "proof" of the theorem of trichotomy (or rather, of the theorem that every cardinal is an *aleph*, from which follows as corollary the trichotomy), which, however – as Zermelo will observe – is not convincing as it appeals, tacitly and intuitively, to some sort of "principle of choice". Indeed, the first correct proof of the theorem, in the context of the explication of the *axiom of choice* from which it depends, is given by Zermelo in 1904.

[301]More correctly, this theorem should be called the theorem of Dedekind–Schröder–Bernstein. Cantor, in fact, obtained it as a corollary of the (never proved) theorem of comparability or trichotomy, until the young Bernstein – in a seminar held around Easter 1897 – gave a demonstration of it, obtained in the previous year, completely independently from the comparability (see the letter to Dedekind of 30 August 1899). Bernstein's proof was relayed by Cantor to E. Borel, who published it in his *Leçons* of 1898. In 1896, independently, also Schröder had tried to give a proof of the theorem of equivalence, but (as Korselt observed in 1911) this attempt contains an error. Finally, also Dedekind had found, already in 1887, a proof of the theorem (in the equivalent form: if $A \subseteq B \subseteq C$ and $A \cong C$ then $B \cong C$): Dedekind told Cantor in his letter of 29 August 1899, but the proof was published only in his 1931.

[302]This is further evidence for our contention that the study under consideration could not have been written in 1891.

[303]Cantor 1895, 484; transl. 90.

1.19.5 Infinite and Finite Numbers, Natural Numbers and Their Classification

As we already mentioned in the Introduction to this Appendix, Husserl insists that a rigorous arithmetical "system" requires the preliminary differentiation of cardinal numbers into *finite* and *infinite*. Beginning with the construction of the natural number series through the operation of 'successor' and then claiming to prove that the series of naturals determines a complete and systematic classification of the field of finite numbers, produces a "non-complete" or "deficient" (*mangelhaft*)" system, since it makes tacit use of a fundamental presupposition: that of finiteness.[304] In other words, it fails to acknowledge that an arithmetic is constructed only for the finite numbers, that the principles and laws that are formulated step by step are valid only for these numbers. Registering this as a deficiency of his *Philosophy of Arithmetic*, Husserl maintains here that the definition of finiteness must be placed at the beginning of the investigations and put into the correct systematical relation with the other definitions. Hence the construction of a rigorous system of arithmetic must be articulated in the following three steps:

(i) Differentiation of finite and infinite cardinal numbers.
(ii) Construction of the natural number series.
(iii) Proof of the fact that all finite numbers have their equivalent in a number of the series.

For (i), we find – preceded by the "standard" example of the one-to-one correspondence of the set of whole positives with the set of odd numbers – Dedekind's definition of infinite set,[305] which is, however, *not formulated for sets in general, but for sets of units, i.e, numbers*:

Definition 1. *"A number is said to be infinite if among its proper parts (unter ihren Teilanzahlen) there is one that is equal to it.*[306] *A number for which this is not true is finite. From its units, therefore, no proper part can be formed that is equal to it."*[307]

Husserl mentions the following formal consequences of Definition 1:

Theorem 1. *The part of an infinite number that is equal to it is itself also infinite.*

Theorem 2. *No infinite number whatsoever could ever be equal to a finite one, i.e. no number can be both finite and infinite.*

---

[304]Tieszen 1990, 153 only alludes to this point without elaborating it any further.

[305]"A system S is said to be *infinite* when it is similar to a proper part of itself; in the contrary case *S* is said to be a *finite* system" (Dedekind 1888, 18, def. n. 64; transl. 63). In Dedekind's terminology, a representation ϕ of a system (or set) S in itself (i.e. a function from S to S) is called *similar* when it is injective, i.e. when to different elements *a*, *b* of system S correspond different ϕ-images.

[306]That is, equivalent.

[307]*PdA* App. 395; *PoA* 369 ff.

Theorem 3. *If a is a finite number, then a + 1 is also a finite number.*

We find a sketch for a proof of Theorem 2 in the *Beilagen.*[308]

Proof. By contradiction, let Z be a finite number and Z′ an infinite number, such that

$$\text{(i) } \mathrm{Z} = \mathrm{Z}', \quad \text{(ii) } \mathrm{Z}' = \vartheta(\mathrm{Z}'),$$

for some proper part $\vartheta(Z')$ of Z′. Then, by the definition of equality among numbers:

$$\text{(iii) } \mathrm{Z} \cong \mathrm{Z}', \quad \text{(iv) } \mathrm{Z}' \cong \vartheta(\mathrm{Z}').$$

It follows from (iii) and from the fact that $\vartheta(Z')$ is properly included in Z′ that $\vartheta(Z) \cong \vartheta(Z')$ for some proper part $\vartheta(Z)$ of Z, but then

$$\mathrm{Z} \cong \mathrm{Z}' \cong \vartheta(\mathrm{Z}') \cong \vartheta(\mathrm{Z}).$$

And therefore Z, being equivalent to one of its proper parts, would be infinite, contrary to the hypothesis. Q.E.D.

The (more complex) proof of Theorem 3 is given in full in the text. It is worthwhile to reconstruct it here in all details and with some additions.

Proof. Let *a* be finite and let us suppose, *per absurdum*, that *a* + 1 would be infinite (keep in mind that here 1 is a unit that *does not belong to the number a*). By the definition of 'infinite number', *a* + 1 can be put in one-to-one correspondence to one of its proper parts, let say $\vartheta(a+1)$: hence

$$f\colon (a+1) \cong \vartheta(a+1) \text{ for a certain correspondence } f.$$

Let Θ be the collection of units of $\vartheta(a+1)$ that is in bijection with *a* (i.e.: the set of the images of the elements of *a* under *f*), while the element 1 of *a* + 1 is correlated with a certain unit 1_0 of $\vartheta(a+1)$ (i.e. $1_0 =_{\mathrm{df}} f(1)$). Hence:

(i) $f\colon a \cong \Theta$; (ii) $\Theta + 1_0 \equiv \vartheta(a+1)$.

If Θ were to contain *only* units of *a* (i.e. if $\Theta \subseteq a$ were to obtain) then, for (i) and the finiteness of *a*, Θ would have to contain also all the units of *a*, given that *a* as a finite number cannot be equivalent to one of its proper parts. Therefore $\Theta \equiv a$ would obtain. But then, 1_0 would coincide with 1: if indeed 1_0 were different from 1, necessarily 1_0 would belong to *a* and, for $\Theta \equiv a$ established before, to Θ. Hence we would have $f(x) = 1_0$ for some $x \in a$, and by the injectivity of *f* and $1_0 =_{\mathrm{df}} f(1)$ it would follow $x = 1$, i.e., $1 \in a$: contradiction. In conclusion, if $\Theta \subseteq a$ were valid, we would have, using (ii):

[308] *PdA* App. 404; *PoA* 379 ff.

(iii) $\vartheta(a + 1) \equiv \Theta + 1_0 \equiv \Theta + 1 \equiv a + 1$

against the fact that $\vartheta(a+1)$ is a *proper* part of $a + 1$. Therefore

(iv) $\neg(\Theta \subseteq a)$

and, since $\Theta \subseteq \vartheta(a+1) \subseteq a+1$,

(v) $1 \in \Theta$.

But then also

(vi) $1_0 \in a$ (because if 1_0 were to coincide with 1 then, as above, we would have that $1 \in a$).

Let us then consider the set of units $\Theta' =_{df} (\Theta - 1) + 1_0$. By (ii), (v) and (vi) we have

(vii) $\Theta' \subseteq a$ (viii) $\Theta' + 1 \equiv \Theta + 1_0 \equiv \vartheta(a + 1)$.

On the other hand, if we adequately modify the correspondence f (by mapping 1 on 1 and the x such that $f(x) = 1$ on 1_0, and by leaving the rest unchanged) we obtain an f' such that:

(ix) $f'\colon \Theta' + 1 \cong \vartheta(a + 1)$.

From (vii), (viii) and (ix), repeating the previously developed argumentation under the assumption that $\Theta \subseteq a$ would obtain, we get that $\vartheta(a + 1) \equiv a + 1$, against the fact that $\vartheta(a + 1)$ is a *proper* part of $a + 1$. Q.E.D.

Husserl now moves on to the second phase of the construction of the system of natural numbers, i.e. to the definition of the *series of natural numbers*.[309] The idea is to proceed by iterative position of the units and parallel collection of the sets of units constituted step by step into a unitary whole. More exactly, once a general sign is established, '1', as indicating "something" (*Etwas*) or "one" (*Eins*), we construct the numbers stepwise, thus:

$$1 + 1, (1 + 1) + 1, ((1 + 1) + 1) + 1, \ldots$$

increasing the previously generated number (i.e. the previously generated *set of units*), at every step, by one unit. We then denote the generated numbers by 2, 3, 4, . . . : "We have therefore the chain of definitions: $2 = 1 + 1, 3 = 2 + 1, 4 = 3 + 1 \ldots$ The series of numbers thus defined we call 'the series of natural numbers'."[310]

[309] About the ordering in a series, Husserl makes a remark that might seem marginal, but that is of great importance from a philosophical point of view: the ordering in a series is not something extrinsical to numerical concepts, but it is *a priori* and intrinsical to the very nature of these concepts. This constitutes "the fundamental fact of arithmetic". Fields of knowledge for which the order of the elements is analogous to that of the numerical field give rise to theories that are "equiform" or potentially identical to arithmetic. The relations and connections among the elements of these fields can be interpreted arithmetically. See *PdA* App. 398–399; *PoA* 373.

[310] *PdA* App. 397; *PoA* 371 ff.

That this would really be a rigorous "definition" is quite debatable. It is clear that natural numbers are all those and only those numbers obtainable starting from *zero* (or *two*, as in this case) using a finite number of iterative additions of a new unit. The problem is that if we want to give a logically sound definition, we would have to do this without any appeal to a "finite number of iterative additions" or, what is the same, without using the expression 'and so on' or the ellipsis '...'. While Frege – as is well-known, and as we will have occasion to recall later on – is clearly and fully aware of this, Husserl does not seem to regard it as problematic, assuming in a certain sense that the correctness of the definition above is warranted by the finiteness of the number $1 + 1$ and by Theorem 3, on the basis of which we know that the operation of *addition of a unit* ('+1', i.e. the successor operation) leads from finite numbers to finite numbers. A further, grave logical omission can be found in the first of the Theorems that Husserl appends to the definition we just encountered, which (together with Theorem 8, see below) one could call '*Theorem of infinity*':

Theorem 4. *The natural number series has a beginning but no end.*

Husserl's alleged *proof* is the this: the first part of the Theorem is trivial; concerning the second, due to axiom [HU.1] every collection (except the total class) can be augmented with a unit, and from this follows that natural numbers do not have an end. Q.E.D. The problem here is: who tells us that if *n* is any number of the series, then *n* is *different from the total class*? [HU.1] is not at all an axiom of infinity!

At this point it is useful to recall – even if only in a schematic outline and without presenting the proofs – how Frege tackled and solved[311] the two issues raised above: the definition of the set of natural numbers and the theorem of infinity.

With every concept F there is associated, according to Frege, a well-determined object, the *number of F* (that we will indicate by *nu(F)*), defined as the *extension* of the concept (of second level) 'concept that is in bijection with F'. Cardinal numbers are then defined as all and only those objects *k* that are the number of *F* for some concept *F*. It is possible to prove (by Basic Law V) *Hume's principle*: $nu(F) = nu(G)$ if and only if *F* and *G* are in one-to-one correspondence, i.e. bijection.

How is it possible to isolate, within the collection of numbers, the natural numbers, i.e. those numbers that can be reached starting from the number *zero* through a *finite number* of steps to the *successor*? This characterization is, obviously, circular: Frege's strategy consists in attempting to define the three concepts

[311] This is the approach already delineated informally in the *Begriffschrift* (1879) and pursued formally in the *Grundgesetze* (1893). As is well-known, the system of the *Grundgesetze* is inconsistent, and hence from this point of view the provability of a theorem of infinity, as that of any other proposition, is obvious. But what is interesting is that the proof – as we find it in the *Grundgesetze* – of the so-called 'Theorem of Frege' (i.e. of the fact that the system of natural numbers as defined by Frege verifies Peano's axioms) can be reproduced in a *consistent* system obtained by substituting *Hume's Principle* (that Frege proves using his Basic Law V, the one responsible for the contradiction) for Basic Law V; cp. Heck 1993; Hale & Wright 2001.

that are involved (zero, the successor relation, being reachable in a finite number of steps) *without presupposing the concept of finite number.*[312]

For the first two notions there are no particular problems. Zero (0) is defined as *nu (F)*, where F is the concept 'object different from itself', i.e.: $Fx \leftrightarrow x \neq x$. The relational concept of successor (S) is defined as the one that subsists between an object x and an object y when there are a concept F and an object a that falls under F, such that $y = nu(F)$ and x is the number that belongs to the concept 'object different from a that falls under F' (one proves that S is functional, i.e. univocal to the right).

The cornerstone of Frege's solution, however, is the successful, non-circular logical characterization of the third notion. In general, given any relation R, we can introduce the exponential powers (the so-called 'peirceians') R^n of R, for n a natural number greater than zero, on the base of the recursive definition

$$R^1 := R$$
$$R^{n+1} := R^n ; R$$

where ; is the operation of *relative product* between binary relations:

$$(R;T)xy \leftrightarrow \exists z(Rxz \wedge Tzy).$$

We then define R^+ as that relation (called *transitive closure* of R) that occurs between an x and a y when for some $n > 0$ we have $R^n xy$. Intuitively, R^n occurs between x and y if and only if y is reachable from x by a chain of exactly n R-steps:

$$x = x_1 R\ x_2\ R \ldots R x_n = y,$$

and hence R^+ occurs between x and y if and only if x is reachable from y by a *finite number* of R-steps.

Frege demonstrates that it is possible to characterize equivalently the relation R^+ (for arbitrary R) *without using the concept of natural number* (at the price of a quantification over concepts, and hence *impredicatively*). After having introduced the property *Er(F,R)* of 'concept F that is *R-hereditary*', as the one that is valid of F and R if and only if F "propagates" along R, i.e. formally:

$$Er(F,R){:}\ = \forall x \forall y(Fx \wedge Rxy \rightarrow Fy),$$

he proves the fundamental result according to which saying that R^+xy obtains is *logically equivalent* to saying that *y falls under every R-hereditary concept G under which x falls*:

$$(*)\, R^+xy \leftrightarrow \forall G(Gx \wedge Er(G,R) \rightarrow Gy).$$

[312] Cp. Simons 2007, 231.

Using the definitions given above of the zero (0) and the successor (S), together with the non-circular reformulation of "reachable in a finite number of successor-steps" given by (*), it is, finally, possible to define the concept NN of natural number:

$$NNx := x = 0 \vee \forall G(G0 \wedge Er(G,S) \rightarrow Gy).$$

Having thus obtained the concept NN, Frege proves that it satisfies Peano's axioms, specifically the axiom of injectivity of the successor (for the other axioms the demonstration is immediate):

$$\forall x\ \forall y\ \forall z\ (NNx \wedge NNy \wedge NNz \wedge Sxz \wedge Syz \rightarrow x = y)$$

which is equivalent to the infinity of NN and hence to Husserl's *Theorem 4* mentioned above. Frege's proof, particularly acute and complex, consists essentially in showing that if x is a natural number (i.e. NNx holds) then, F being such that $nu(F) = x$ and v being the extension of the *total concept* V ($Vx \leftrightarrow x = x$), we obtain that v *does not fall* under F.

After this digression on Frege, we conclude with the presentation of the remaining five theorems in Husserl's text, which aim at establishing the main result according to which *for every finite number one and only one representative is included in the series* (Theorem 9).

Theorem 5. *1 + 1 is a finite number.*

Theorem 6. *The natural number series contains finite numbers only.*

This follows from the "definition" of the numerical series (i.e. "by induction") by Theorems 3 and 5.

Theorem 7. *The number series contains pairwise unequal numbers only.*

Proof. For the construction of the series, if the number Z_1 follows in the order of the series after the number Z_0, then all the units of Z_0 are contained in Z_1 but not the other way around (the unit that is added with every step is *new*), i.e. $Z_0 \subset Z_1$. Hence, if we had $Z_0 = Z_1$, then Z_1 would be equal to, i.e. in bijection with, one of its proper parts, and hence it would be infinite, contrary to Theorem 6. Q.E.D.

Theorem 8. *The number of the natural numbers is an infinite one.*

Proof. The set of odd numbers, generated starting from the unit 1 by iteratively adding two new units, is a succession that is a proper part of the natural one, and that can be put in bijection with it. Q.E.D.

With Husserl's last Theorem, we finally arrive at the conclusion that the natural number series allows a *complete* classification of the field of finite numbers:

Theorem 9. *To each finite number there corresponds in the series of natural numbers one, but also only one, equal to it.*

Proof. Let a be a certain finite number. Take any pair of units of a and form the natural number $1 + 1$ (=2). Then we proceed – until exhausting all the units of a – to extract from a one unit at a time, adding it to the number of the series we have obtained at the previous step and obtaining thus the numbers $3 = 2 + 1, 4 = 3 + 1$, etc. (in other words: we count a). The natural number resulting when a has been "emptied" of all units is equal to (i.e. in bijection with) a, and it is certain that this process leads to a final natural number, because otherwise the infinite natural series (Theorem 8) would be in bijection with the finite number a, which is impossible. Moreover, if there were more than one natural number equal to a, there would be two different natural numbers that would be equal among themselves, which is incompatible with Theorem 7. Q.E.D.

Husserl emphasizes that in the proof of Theorem 9 the order in which we extract the units from a is indifferent: in whatever order, the resulting number of the series is always the same.[313] In other words, it is not necessary to prove beforehand a Lemma that states that if two finite sets are in bijection with respect to a certain order of their elements, then they are in bijection with respect to any ordering, since it follows from Theorem 9.[314]

1.20 Concluding Remarks

In this chapter we have pursued four different goals: (i) a precise positioning of early Husserl's work in the field of logic and the philosophy of mathematics; (ii) a mathematical specification of some of his intuitions by means of standard logical notions and tools wherever possible; (iii) a comparison with Frege concerning certain very specific topics such as the *use of defining a concept by defining its extension* or the definition of cardinal numbers (cp. Appendix 1); (iv) a comparison with the algebraists of logic, in particular Boole, as to the treatment of the symbolic aspect of the algorithmic systems (cp. Appendix 2).

Husserl came to the fore as a mathematician who confronts the problems of the mathematical world in which Weierstrass and Cantor were active, reflects on the issue of the *arithmetization of analysis*, appropriates Cantor's definition of cardinal numbers through abstraction and reflects on Cantor's *Mannigfaltigkeitslehre*. He fully understands the demands to which the critical movement devoted to the

313 Also see *PdA* 109; *PoA* 114.

314 Here Husserl seems to answer one of the objections that Frege made in his Review of the *Philosophy of Arithmetic*, i.e. that if one maintains that the fastest way to compare the cardinal numbers of two sets is counting the elements of each set, and furthermore, if one maintains that the actual reason for which two sets turn out to be in bijection is that they have the same number and not the other way around, one commits the error of neglecting that "counting the elements" means precisely to put them in bijection with a segment of the series of naturals.

"instillation of rigor in analysis"[315] tries to give an answer as well as the tendency towards formalization and generalization present in the mathematics of the nineteenth century. From his reception of the results of abstract algebra and his conception of algorithmic systems as capable of applying to diverse systems of entities, it emerges that he "clearly endorses a limited version of the thesis later to be defended by Wittgenstein and Gentzen of meaning as use."[316] Especially important in Husserl's youthful work are his reflections on arithmetical operations; in particular his inquiry into the possibility of circumscribing the 'class of all conceivable arithmetical operations' and his claim that the addition operation is irreducible to that of multiplication.[317] We believe that the attempt at finding a formal counterpart for the class of functions Husserl was thinking of can be viewed as an example of how his intuitions can be developed and completed by *argumenta in forma*. One of the consequences of this choice is that it makes possible to see Husserl as a pioneer in his attempt to investigate the question whether the class of "*computable* arithmetical functions" can be characterized *as a whole*. The problem of finding an adequate *mathematical* characterization (or *definition*) of the *intuitive* notion of computability has received various answers much later (since 1933–34), and, to the best of our knowledge, nobody has ever thought of Husserl as someone who has anything to contribute to this issue. On the whole, a formal reconstruction can help to overcome the regrettable unclarity of many of Husserl's formulations. From the perspective of the history of logic the specific proof of the equivalence of Husserl's class of functions with the class of functions known in current logic as the class of partial recursive functions represents a substantially new result.

[315]Kline 1972, 947–978.

[316]Mulligan 2004 (unpublished).

[317]See §15.

Chapter 2
The Idea of Pure Logic

2.1 Introduction

In his *Prolegomena to Pure Logic* Husserl works with Bolzano's idea that the entire field of truths can be partitioned into several parts, each of which consists of all truths "of a certain kind," that is, all truths that are germane to a certain homogeneous kind (*Gattung*) of objects.[1] Husserl says that "it is not arbitrary where and how we delimit fields of truth,"[2] "the domain of truth is not an unordered chaos,"[3] but it is articulated in "natural provinces"[4] that are also called "fields of knowledge (*Erkenntnisgebiete*)"[5] or fields of experience (in a broad sense of this word which comports well with common mathematical usage, where, for instance, a system of abstract objects like e.g. the natural numbers equipped with certain functions and relations is said to be a "field of experience"). Each field of experience in Husserl's sense can be viewed as "an independent reality with its own experimentally determined mathematical structure."[6] In this sense fields of "the purely mathematical sciences whose objects are numbers, manifolds (*Mannigfaltigkeiten*), etc., things thought of as mere bearers of ideal properties, independently from real being or not being,"[7] are also to be considered fields of knowledge or of experience.[8]

---

[1]Cp. the allusion to Bolzano's *WL* in *PR* 29, *PRe* 73.

[2]*PR* 5, *PRe* 54.

[3]*PR* 15, *PRe* 62.

[4]*PR* 25, *PRe* 70.

[5]*PR* 19, *PRe* 65.

[6]Here I borrow the terminology from Webb 1980, 79.

[7]*PR* 11, *PRe* 69.

[8]Tieszen 2004 rightly stresses that "Husserl also says that among the eidetic sciences some are exact and some are inexact. Mathematics and logic are exact ... mathematics and logic set the standard for what is clear, distinct and precise" (33–34). Our considerations in this chapter always refer (unless otherwise specified) to exact sciences.

S. Centrone, *Logic and Philosophy of Mathematics in the Early Husserl*, Synthese Library 345,
DOI 10.1007/978-90-481-3246-1_2,

Truths that constitute a "natural province" are organized according to a certain objective relation of dependency.[9] It is the task of each single science to make manifest this relation of dependency among the truths concerning its own field of objects.

Seen from this perspective, the sciences are characterized by methodologies apt to transmit cognitive evidence – with respect to circumscribed regions of the world – ranging from those facts (obtaining states of affairs) which can be directly recognized as facts, to those for which this is not possible. Within each single science the transmission of evidence occurs through certain typical "ways of proceeding" (*Verfahrungsweisen*) common to all disciplines generally accepted as such (arithmetic, geometry, natural sciences). Husserl reserves the name of "groundings" or "foundations (*Begründungen*)"[10] for these "typical ways of proceeding."[11] These are *classes of typical inferences* (e.g., the syllogism in *Barbara*), that is, all and only those used in the actually existing sciences. The form of the science is determined by the connection (*Verbindung*) and the order (*Ordnung*) of the foundations. They constitute what Husserl refers to as a "systematic interconnection in the theoretical sense (*systematischer Zusammenhang im theoretischen Sinne*)".[12]

Three different connotations of the term 'logic' can be found in the *Prolegomena*. Logic is conceived of, firstly, as *Wissenschaftslehre* in the sense of a theory of scientific methodologies. Thus understood, it assumes effectively the traits of a metalogic.[13] Secondly, *logic* is intended as *pure logic* which studies relations between certain abstract non-linguistic entities, such as concepts and propositions.

Used in the former sense, the term "logic" refers to a meta-theoretical discourse on scientific theories: as "systematical webs of groundings (*systematische Gewebe von Begründungen*)"[14] they become the object of the investigation. This is a

[9]Bolzano 1810: "In the realm of truth, that is in the collection of all true judgements, reigns an objective interconnection that is independent of the contingent fact that we subjectively acknowledge it; it is in virtue of this that some of those judgements are the reasons of others and the latter the consequences of the former" (Part II, §2). Cp. Cavaillès 1938, 54–55.

[10]It would be more correct to translate "*Begründung*" as "grounding" and to reserve "foundation" for "*Fundierung*". The former is Husserl's version of Bolzano's "*Abfolge*", as we shall try to show. The latter is used in the third *LU* to signify one of the possible dependence relations between the parts of an object. However, for the sake of fluency of style we shall use both terms (accompanied by the German word in brackets).

[11]See for instance the title of §9 (*PR* 22, *PRe* 68): "Methodical ways of proceeding in the sciences – in part groundings, in part auxiliary devices towards groundings".

[12]*PR* 15, *PRe* 62.

[13]"The task of the theory of science will therefore also be to deal with the sciences as *systematic unities of this or that sort...*" Each science "can be subsumed under the concept of method, so that the *Wissenschaftslehre*'s task is not merely to deal with the methods of knowledge in the sciences, but also with such methods as are themselves styled sciences" (*PR* 25, *PRe* 70).

[14]*PR* 25, *PRe* 70.

broadening of Bolzano's concept of a *Wissenschaftslehre*,[15] "a normative and practical discipline relating to the Idea of science,"[16] as Husserl puts it. Here 'practical' is meant in the Bolzanian sense: a science that "also has the task of formulating the rules according to which the sciences must be delimited and constituted."[17]

As regards the second sense of "logic", it is well known that Husserl's refutation of logical psychologism leads in the *Prolegomena* to the identification of "an internally closed, independent ... field"[18] of a priori truths, which constitute the domain of pure logic. Pure logic acknowledges the objectivity of contents of thinking (concepts, propositions, inferences) and studies the properties of and the logical relationships among them. It is a formal, theoretical, a priori science, *independent* of other sciences, and, in particular, of psychology. Here 'independent' is meant in Bolzano's sense: in order to prove its derived propositions (theorems), pure logic does not require any auxiliary truths that are not logical truths. According to Bolzano, a science A is dependent (*abhängig*) on another science B (takes from another science part of its own theoretical content, as Husserl puts it[19]) if in a textbook of A some truths of B are indispensable as lemmata (*Hülfssätze*, auxiliary truths) for proving the theorems of A.

> For example, the theory of space (geometry) depends in this fashion on the general theory of magnitudes (arithmetic, analysis) because textbooks of the former contain certain truths, indispensable for the proofs of some of its essential theorems, which deal with magnitudes in general rather than with space in particular and hence essentially belong to the general theory of magnitudes.[20]

A science is "altogether independent" just in case it is not dependent on any other science. So Husserl takes the field of the truths of logic in the second sense to be the domain of an *independent* theoretical science in Bolzano's sense.

Logic is thirdly thought of as a *Wissenschaftslehre* in the sense of a theory concerned with the *deductive mechanism* in general. In this sense, too, it can be seen

---

[15]A broadening insofar as, besides the problems of order, organization and systematization which pertain to the exposition (*Darstellung*) of a theory, we find in Husserl, from c.1896 onwards, the idea that not only mathematical but all theories insofar as they are formalized are to be made the object of the investigation. And formalization in this sense is not present in Bolzano. For more on this issue see next chapter.

[16]*PR* 12, *PRe* 60.

[17]*PR* 29, *PRe* 73. Actually Bolzano defines *Wissenschaftslehre* as "the aggregate (*Inbegriff*) of all those rules which we must follow when subdividing the entire realm of truth into single sciences and representing them in special textbooks, if we want to proceed in a useful way" (*WL* I, §1, 7).

[18]*PR* 32, *PRe* 76.

[19]See for instance *PR* 47, *PRe* 87: "it is ... easy to see that each normative, and, *a fortiori*, each practical discipline, presupposes one or more theoretical disciplines as its foundations, in the sense namely, that it must have a theoretical content free from all normativity, which as such has its natural location in certain theoretical sciences ..."

[20]*WL* I, §13, 53.

as a "science of science",[21] but now it is taken to be concerned with the logical mechanism as a feature of all formal sciences. Thus conceived it consists of laws that formally warrant the step from axioms to theorems.[22] In the last chapter of the *Prolegomena* Husserl speaks of "a sphere of laws, which in formal universality span all possible meanings and objects, under which every particular theory or science is ranged, which it must obey if it is to be valid".[23] These are the laws with which every formal theory is to comply "and through which, as a theory validated by its form, it can be ultimately justified".[24]

2.2 The Concept of a Theory

As we already pointed out in the Introduction to this chapter, from 1896 onwards Husserl works with an essentially Bolzanian conception of a theory as a collection of true propositions about a certain sphere of objects, which are either primary and indemonstrable principles (*erste Grundsätze*) or derived from such principles in accordance with a fixed set of rules. It is important to remember that the objects the theory is concerned with must all belong to the same homogeneous kind and that the theory must contemplate *all* the consequences that can be derived from its principles in accordance with the rules.[25] This conception of a theory does not appear to be substantially different from what emerges from the *classical* idea of an axiomatic system that was outlined by Aristotle in his *Analytica Posteriora* and of which Euclid's *Elements* are the most representative example.[26] In this light, a theory is a tool that enables knowledge about a clearly demarcated realm of objects to be *organized systematically*. In order to be *categorized* a certain limited number of *primitive concepts* must be identified that are both *immediately intelligible*, hence not in need of definition, and *sufficient* for every other concept pertinent to the field to be related to them by means of the logical tool of *definition* and thus to acquire, albeit indirectly, intelligibility. As regards *the description of the properties* of the field, a certain limited number of *primitive propositions* (the axioms of the theory) is to be identified, the *truth* of which is *immediately obvious*, hence not in need of proof, and which *suffice* for the derivation of every other true proposition about the field by means of the logical tool of *demonstration*. The foundations of the theoretical construct – primitive concepts and primitive truths – thus rest on something *extra-logical*, that is on the immediate comprehensibility of some concepts and on the self-evidence of some propositions. The development of the construct,

---

[21]*PR* 12, *PRe* 60.

[22]Cp. Tieszen 2004, 28.

[23]*PR* 246, *PRe* 239.

[24]Loc. cit.

[25]For this conception see also Jan Berg, BGA, vol. 11/1, 18.

[26]The reference paper on this classical view is Scholz 1930; cp. also Casari 1973.

by contrast, occurs by means of the *logical* procedures of definition and demonstration that endow defined concepts with intelligibility and derived propositions with truth.

What, if anything, distinguishes the conception of a theory shared by Husserl and Bolzano from the view just characterized? At this point some conceptual distinctions made by Casari turn out to be very helpful. He calls the conception of a theory of which the view just described is one variant a *categorical* conception of an axiomatic theory. Now "with respect to the distinction between primitive and derived or derivable, this categorical conception gives ... rise to two variants that we can distinguish as the *epistemic* and the *etiological* conception."[27] The contrast is manifest already at the level of proofs: "... from the epistemic perspective a proof is a procedure through which we *ascertain* the truth of an assertion, while from the etiological point of view a proof is a procedure by which we *bring to light the reasons* of the truth of the assertion."[28] At the level of a theory T this contrast reappears as the distinction between the *various* possible logical presentations of T that are epistemically adequate (insofar as every non-axiomatic truth of T is provable in them) and a *privileged* presentation of T in which every non-axiomatic truth of T is "etiologically proven". Can one find some suitable formal conditions by which such a privileged logical presentation of T can be isolated (even though the content of the relevant propositions cannot be set aside completely)? The most important formal condition is the following: the proofs must *proceed from the simpler to the more complex*, in conformity with the intuitive requirement that in a good explanation the *explicans* must not be more complex than the *explicandum*. Thus, at the level of invertible logical inferences (such as "A, B, therefore $A \wedge B$" and its inversions "$A \wedge B$, therefore A", "$A \wedge B$, therefore B"), the direction of introduction is privileged. Another formal condition and, intuitively equally justified, is this one on which Bolzano also dwells: *etiologically acceptable proofs must proceed from the general to the particular.*

Bolzano points out that the contrast between the epistemic and the etiological conception of proof largely coincides with the distinction, marked by Aristotle and the Scholastics, between those proofs that simply show *that* something is the case (*demonstratio quia*) and those that explain *why* something is the case (*demonstratio propter quid*), which give the objective reason for its being the case.[29] He also captures the contrast as that between proof as *ascertainment* (*Gewißmachung*), which aims at producing certainty as regards the proposition that is to be proven, and proof as *foundation* or *grounding* (*Begründung*), which rather aims at giving the reasons for the proposition in question.[30] Bolzano had insisted on this distinction between "subjective" and "objective proofs" already in 1810:

---

[27]Casari 1987, 330. The epithet "etiological" alludes to the Greek word '*aitía*' which means whatever is specified in an answer to a *why*-question. (Aristotle's famous theory of the "four *aitíai*" is a theory of four kinds of *because*, rather than a theory of four kinds of cause.)

[28]Op. cit., 331–332.

[29]Bolzano, *WL* II, §198, 341. Cp. Aristotle, *An. Post.* I, 13; Aquinas, *Summa Theologiae* I, quaestio 2, art. 2.

[30]Bolzano, *WL* IV, §525, 261–2. Cp. Bolzano 1834, I, §3, No. 2.

> by a scientific proof of a truth we must understand the presentation of the objective dependence (*objective Abhängigkeit*) of a given truth on other truths, i.e. the deduction of a truth ... from such truths as must in themselves and necessarily be regarded as its ground.[31]

As can be seen from *Formal and Transcendental Logic*, Husserl knew Bolzano's early monograph.[32]

Unlike the traditional idea of a rigorous deductive science (viz., the categorical conception of an axiomatic system in its *epistemic* variant) the concept of theory with which both Husserl and Bolzano operate (viz., the categorical *etiological* conception of such an axiomatic system) needs an account of the nature of *dependency relations* among true propositions. Bolzano explicitly requires that in the process of backtracking from a truth to its reasons we want to find neither (i) "extraneous material" with respect to the conclusion nor (ii) principles that are more specific than the truth to be proven, that is we want a proof of a given truth that moves from simplest to most complex and from most general to most specific. Husserl accepts, as we shall see, both these formal conditions.

In the next sections we will try to do three things: (a) clarify the Husserlian concept of *foundation* or *grounding* (*Begründung*) and its relation to the Bolzanian concepts of *derivability* (*Ableitbarkeit*) and *consecutivity* (*Abfolge*), (b) elucidate the notions of "*interconnection of things*" and "*interconnection of truths*," as used in Chapter XI of the *Prolegomena*, and (c) explain what Husserl means when he says that a concept (*Begriff*) founds or determines a conceptual field (*Begriffsgebiet*).

2.3 The Concept of *Begründung*

As we have seen, in the *Prolegomena* Husserl works with Bolzano's idea that there is *a certain objective connection* among truths, independent of the cognitive activity of the subject: certain truths are the "reasons" (*Gründe*) of others and the latter are "consequences" (*Folgen*) of the former. Husserl characterizes this relation between truths as "a certain objective or ideal interconnection which provides [certain cognitive acts ore states] with a common objectual reference (*gegenständliche Beziehung*) and thereby also with ideal validity (*Geltung*)".[33] In this "systematic interconnection in the theoretical sense (*systematischer Zusammenhang im theoretischen Sinne*) ... lies the foundation of knowledge (*die Begründung des*

[31] 17, II, §12; cp. Bolzano 1834, I, §3, No. 2.

[32] Op. cit., §26 d). The editor of this volume of the Husserliana neither gets the title of Bolzano's booklet right nor the name of the editor of the 1926 edition (Heinrich Fels). Husserl says that this early work of Bolzano had been "nearly unavailable". This does not imply that he himself only came across it in 1926. Another Brentanist, Benno Kerry, had referred to it already in the eighties of the nineteenth century: see Künne 2009, 327.

[33] *PR* 228; *PRe* 225.

Wissens) and suitable connection and order in the sequence of those groundings."[34] That is to say, the transition from truths that are neither capable nor in need of a proof because they are self-evident (principles/axioms, *Grundsätze/Axiome*) to other truths that require a proof is effected by means of *Begründungen*, that is it is an etiologically acceptable proof.

In his *Wissenschaftslehre*, Bolzano introduces two important relations among propositions: *consecutivity* (*Abfolge*), a relation that can obtain only between true propositions and that connects certain truths as reasons to others as their consequences, and *derivability* (*Ableitbarkeit*), a *formal*[35] relation that does not only obtain between true propositions and that corresponds to the concept of "following from certain premises through correct inferences." The relation of derivability is characterized by Bolzano in such a way that it covers not only cases of formal inference, but also cases of semiformal enthymematic inference.[36]

Because it can obtain only among truths, the relation of *Abfolge* is not purely formal: it also depends on the content of the propositions between which it obtains. More exactly, *Abfolge* is generally speaking a material relation. However, Bolzano seems to think that the *Abfolge* relation confined to *conceptual* truths is as a formal relation. In this case the question of the consecutivity among true conceptual propositions can be recast as the question whether the steps of a proof obey a principle of analyticity, that is, whether the premises of the inferences do not contain concepts or propositions that are not contained in the conclusion. This is a condition which can be fruitfully related to that of a proof in normal form that is central in contemporary proof theory.

We maintain that the Husserlian concept of *Begründung* is a direct adaptation of the Bolzanian notion of "*consecutivity* between truths."

First of all, we have to point out that in Husserl's characterization of the scientific process as interconnection of *Begründungen* the concept of logical inference is privileged. For him the truly important concept in deductions is "logical inference", though Bolzano, as we shall see,[37] is less certain that each case of consecutivity is a case of logical inference.

"Scientific knowledge", Husserl claims, "is as such grounded knowledge (*Erkenntnis aus dem Grunde*)."[38] He shares with Bolzano the view that the rules of inferences that give the reasons for a truth must be analytic (in Gentzen's sense) and intrinsic: the premises must be simpler than the conclusion, and they must not contain material that is extraneous to the latter. By contrast, showing that a certain proposition A is a logical consequence of a certain proposition B is obtaining A from B through correct inferences. The relation of logical consequence

---

[34]*PR* 15; *PRe* 62.

[35]The relation of derivability is characterized by rules of inference that exclusively concern the form of the involved propositions.

[36]See Appendix 4 to this chapter for some explanation of these notions.

[37]See below.

[38]*PR*, §63, 231.

(corresponding to Bolzano's derivability) is, in this sense, a weaker relation than that of grounding (*Begründung im prägnanten Sinne*),[39] which is Bolzano's *Abfolge* by another name. We can maintain with reasonable certainty[40] that Husserl's position on the relation between "foundations" or "groundings" and valid inferences is as follows: *not all valid inferences are foundations* (yield the "why", the *Grund*), *but all foundations are valid inferences*. Thus, for example,

$$\mathrm{A, B, therefore\ A} \land \mathrm{B}$$

is both a valid inference and a foundation, whereas

$$\mathrm{A} \land \mathrm{B, therefore\ A}$$

is a valid inference but not a foundation. Though A can clearly be logically inferred from A ∧ B, one cannot sensibly maintain that A ∧ B constitutes the (or a) reason of A. The following passage is very telling in this respect:

> Notice the following distinction: every explanatory (*erklärende*) interconnection is a deductive one, but not every deductive interconnection is an explanatory one. All reasons (*Gründe*) are premises, but not all premises are reasons. Every deduction is necessary, i.e. falls under laws, but the fact that the conclusions follow *according to* laws (inferential laws) does not mean that they follow *from* those laws and are "founded" *("gründen")* in them in an emphatic sense. Of course habitually we refer to every premise ... as "reason" of the "consequence" that is drawn from it – an equivocation that we need to heed carefully.[41]

The first generalization marks a difference between Husserl and Bolzano that is to be registered: Bolzano is by no means certain that every case of *Abfolge* is also a case *Ableitbarkeit*.[42]

Incidentally, the point Husserl makes in the penultimate statement of the above quotation is also a Bolzanian one. Anticipating by four decades the insight that was unforgettably expressed in Lewis Carroll's "What the Tortoise Said to Achilles", Bolzano clearly distinguished in his *Wissenschaftslehre* between the premises of a deduction and the rules in accordance with which an inference proceeds:

> If one maintains that the complete ground of the truths M, N. O, ... includes, besides truths A, B, C, D, ... , from which they are derivable, also the rule which allows [their] derivation, then this amounts to maintaining that propositions M, N, O, ... are true only because this rule of inference is valid and because propositions A, B, C, D, ... are true. This is tantamount to [another] inference... But since every inference has a rule, this [new inference] does too... We can see at once that this way of inferring can be repeated *ad infinitum*, and that, if it were legitimate to add *one* rule of inference to the ground of the truths M, N, O, ..., an infinite number of them could be claimed to belong to this ground; which seems absurd.[43]

---

[39] *PR*, §64, 233; *PRe* 229.

[40] In this respect, see the following quotation from *PR*, Chapter XI.

[41] *PR* 235; *PRe* 229.

[42] Bolzano, *WL* II, §200, 346–48. But cp. *WL* II, §221 note, 388.

[43] Bolzano, *WL* II, §199, 345. For further references see Künne 2008, 400.

Moreover, what is grounded cannot be more general than its ground. This condition, already put forward by Bolzano,[44] is interpreted by Husserl in a rather special sense. He has in mind here the fact that axioms function as schemata whose instances are used in proofs. This requirement was explicitly stated few years earlier, in the *LV96*, as one of the four principles that must precede the constitution of a deductive theory.[45] There he claims that in every deductive theory the proof of a proposition must consist in subsuming it under more general propositions, that is finally under a subset of the primitive axioms of the theory. He calls this way of proceeding "subsumption under the axioms (*Subsumtion unter die Axiome*)": We do not admit as conclusion "any proposition that does not fall under the basic laws (*Grundgesetze*)."[46] In the *Prolegomena* he writes: "If we are dealing with the grounding (*Begründung*) . . . of a *general* truth, . . . we are referred to certain general laws, which, by way of specialization (not individualization) and deductive consequence yield the proposition to be proved."[47] That is to say, we have to show that the intended proposition is an instance of those laws fixed as basic laws or is obtained from them (or from derived proposition) by means of an inference rule of the kind *Abfolge*.

In §7 of Chapter I of the *Prolegomena* three peculiarities are ascribed to foundations (*Begründungen*). Firstly, they are said to be "fixed structures (*feste Gefüge*)", that is they are valid inferences (the conclusion follows necessarily from the premises), *modulo* making intended premises explicit. Husserl allows for enthymematic inferences, as did Bolzano. In *LV96*, 234 he maintains that to every inference, of the kind later called "grounding" or "foundation", corresponds a *Kausalsatz*, that is a proposition of the form 'A, because B' (where 'A' represents the conclusion of the inference and 'B' the conjunction of its premises); conversely, to every *Kausalsatz* corresponds an inference, not in the strict sense, but *in the broadest sense* ("*wenn der Terminus Schluß in seinem weitesten Sinn genommen wird*"), that is as covering also enthymematic inferences. In Husserl's example, 'Caius is mortal because he is a man' corresponds to the enthymematic inference that concludes 'Caius is mortal' from 'Caius is a man', an inference that, making the tacit premise 'all men are mortal' explicit, becomes an inference in the strict sense.

Secondly, foundations (*Begründungen*) are said to exemplify kinds, or schemata, of inference (*Schlußarten, Schlußformen*): all inferences used in the various scientific disciplines are instances of a finite number of *Schlußarten,* and the problem becomes that of determining *which* and *how many* kinds of simple inference there are.[48] And finally, Husserl emphasizes that no form of *Begründung* is reserved for a special field of knowledge.

---

[44]See below, Appendix 4.

[45]There is a formal treatment of these notions in Appendix 4.

[46]For example, see *LV96*, 246.

[47]*PR* 232; *PRe* 228.

[48]See footnote 49 below.

Although Husserl conceives of scientific proofs as made up of groundings in the emphatic (or etiological) sense (*Begründungen im prägnanten Sinne*), he seems not to object to valid inferences of a less demanding kind in the sciences, that is to proofs that only comply with the demands of *Ableitbarkeit*.[49] Presumably he thereby wants to remain faithful to actual mathematical practice in which derivations that are not of the canonical kind, that is that are not etiological, are liberally used. Maybe he thought that all scientific proofs of this inferior type could be transformed into proofs that are of the canonical kind (although, as far as we know, he made no attempt at proving that this is possible).

There is a further feature of Husserl's views on foundations or groundings (*Begründungen*) that ought to be mentioned: he requires that the move from axioms to theorems is made by inferential steps that are simple. An inferential step is simple just in case it is not possible to decompose the passage from the premises to the conclusion any further. This further requirement is not explicitly formulated by Husserl, but it clearly emerges from reflections upon the second of the peculiarities ascribed to foundations, that is their being inference schemata. We are looking for a possibly limited number of atomic inferences into which all other (more complex) inferences can be decomposed.[50]

Of groundings Husserl says that "science can never do without this helpful ladder (*Stufenleiter*)."[51] We need them "in order to pass beyond what, in knowledge, is immediately and therefore trivially evident." After all, "... evidence ... is in fact only immediately felt in the case of a fairly limited group of primitive facts. Countless true propositions are only grasped by us as true when we methodically ground them." In other words, "there are infinitely many truths which could never be transformed into knowledge without such methodical procedures." The "foundational interconnection (*Begründungszusammenhang*)" characterizes science as such.

[49]See for instance the calculus of an axiomatic-synthetic kind presented in *LV*96 (Appendix 5). One of the principles that must precede the constitution of every deductive theory is the *modus ponens* of traditional logic, that is, the rule: A, A → B/B. This is a typical inference rule of the kind *Ableitbarkeit*.

[50]The inference: a → b, b → c, c → d/a → d can for instance be decomposed into the simpler inferences: a → b, b → c/a → c and a → c, c → d/a → d. That Husserl thinks this way clearly emerges, as we said above, from his characterization of *Begründungen* as schemata "If (the *Begründungen*) were formless and lawless, if it were not a fundamental truth that all *Begründungen* have certain inherent 'forms', not peculiar to the inference set before us *hic et nunc*, but typical of a whole class of inference ..." (*PR* 20; *PRe* 66). Here it is quite natural to think that with these 'forms of inference' that act as patterns for a whole class of particular inferences we are looking for forms of inferences of the most elementary kind.

[51]All quotations in this paragraph are from *PR* 15–16; *PRe* 62–63.

2.4 The Interconnection of Things and the Interconnection of Truths

Husserl introduces an important distinction when he writes:

> Two meanings can be attached to this objective interconnection which ideally pervades scientific thought and which gives 'unity' to science as such: it can be understood as the *interconnection of things* which our (actual or possible) acts of thinking are intentionally directed at, or, on the other hand, as the *interconnection of truths* in which this unity of things acquires objective validity (*Geltung*) as being what it is.[52]

Perhaps the best way to explain this dichotomy is *via* elucidating the sense of the statement that "a concept (*Begriff*) determines a field."[53] Consider the concept of cardinal number (*Anzahl*). The assertion that "the concept of cardinal number determines a field of knowledge" is to be understood as follows: the field under consideration is constituted by the objects that fall under (*unterstehen*) this concept, that is by *infimae species* of the genus 'cardinal number'[54]: they are precisely the cardinal numbers, for which there is a uniform and univocal principle of construction (sc. passage to the successor: $1; 2 = 1 + 1; 3 = 2 + 1; \ldots$[55]). The only relation that obtains between the objects of this field, the natural number series, is the order relation ($\leq$: given any two finite cardinal numbers, either one is smaller than the other or they are the same). The connections "that have . . . a meaning for the objects of this field"[56] are the arithmetical operations (elementary operations: addition, subtraction, multiplication, division, and higher operations: exponentiation, root extraction, logarithm, etc.).

What, then, is the sense of Husserl's often repeated claim that some laws are "grounded in the essence of this or that",[57] in this case: grounded in the constitutive concept of the field under consideration? A law is grounded in a concept if and only if it is *analytically included in that concept*. So the laws of commutativity and

---

52*PR* 228; *PRe* 225.

53Cp., for instance, this passage: ". . .whether it is the domain of cardinal numbers (*Anzahlengebiet*) or some other conceptual domain (*Begriffsgebiet*), that the general arithmetic . . . governs" (*PoA* 7; *PdA* 7). The notion of a "conceptual determination (*Begriffsbestimmung*) that delimits a field of knowledge" derives from Bolzano. See e.g., *WL* I, § 2, 9: ". . . the field of a science that we obtain by means of this conceptual determination (*das Gebiet der Wissenschaft, die wir durch diese Begriffsbestimmung erhalten*)".

54*PoA* 414 ff. *PdA* 434: "If we understand by *cardinal number* the answer to the question 'How many?' then the number series is the closed manifold of *particularizations* that are possible in the sphere of the concept *how many*."

55*PoA* 238 ff., *PdA* 226 ff.

56*PoA* 412, *PdA* 433.

57Cp. for instance *PoA* 414, *PdA* 435: "operations which are *grounded* in the Idea of the cardinal number" or *LV96*, 241: "the general logical laws divide in several groups: in laws which have their roots in the concept of proposition, in laws which have their roots in the concept of concept, in laws which have their roots in the concept of object".

associativity, for example, are grounded in the concept of addition by being analytically included in this concept.[58]

What is the interconnection of things? In this specific case it is the numerical field, organized in the form of a number system: "the system peculiar to science . . . is not our own invention, but is present in things, where we simply find or discover it."[59]

And what is the interconnection of truths? In our example, it is the arithmetic of finite cardinal numbers. The point is that the interconnection of things and the interconnection of truths are, in a way, the same thing, considered from two different perspectives. We have knowledge of a certain field only through the truths holding of that field; so, sticking with our example, we have knowledge of the connections (+, − . . .) and the relations (<, =) holding in the numerical field – and we are able to operate with them – only by knowing the mathematical laws that govern them. Quite generally, Husserl maintains:

> [the interconnection of things and the interconnection of truths] are given together *a priori* and are mutually inseparable. Nothing can *be* without being somehow determined, and that something is and is somehow determined is the truth in itself (*Wahrheit an sich*) which is the necessary correlate of the being in itself (*Sein an sich*). What holds of single truths . . . plainly also holds of interconnections of truths . . . This self-evident inseparability is not, however, identity. In these truths or interconnections of truths the actual existence of things and of interconnections of things finds expression. But the interconnections of truths differ from the interconnections of things which are 'truly' [represented] in the former; this is immediately evidenced by the fact that truths which hold of truths do not coincide with truths that hold of the things posited in such truths.[60]

Thus, for example, no truth that belongs to the arithmetic of natural numbers (elementary number theory) *coincides* with any truths *about* arithmetical truths (like that concerning the incompleteness of the formal system of *PA*).

2.5 The Idea of Pure Logic

Let us now consider Husserl's strikingly innovative view of the structure of formal logic, as it is outlined in Chapter XI of the *Prolegomena*, and then (in the next section) reflect on some of the disciplines that are located at each level in this

---

[58]Fine 1994 maintains that as regards explaining the concept of essence the classical doctrine of real definitions is superior to an account in terms of necessity. Every general law that affirms an essential relation among the objects of a certain field is a necessary truth, but essence cannot be explained by modal notions. Fine appeals to a conception of essence that is admittedly inspired by the Husserlian one, that is the notion of nature/essence of an object is primitive. "[E]ach class of objects, be they concepts or individuals or entities of some other kind, will give rise to its own domain of necessary truths, *the truths which flow from the nature of the objects in question . . .*" An important development of this conception of "to be grounded in the essence of" can be found in Mulligan 2004.

[59]*PR* 15; *PRe* 62.

[60]*PR* 228; *PRe* 225–226.

structure, using for this purpose the sketches of a logical morphology and of a logic of non-contradiction in the fourth *Investigation*.

Husserl characterizes the structure and the tasks of pure logic in the course of answering the question, "what are the conditions of the possibility of a theory in general?",[61] in other words, "what are the primitive ... concepts that constitute the concept of a theory?"[62] Since a theory is made up of "truths (that are articulated) in the purely ideal forms of reason (*Grund*) and consequence (*Folge*),"[63] where truths are considered as true propositions *à la* Bolzano, the science whose task it is to elucidate the structure of theories will also have the task to isolate the kinds of notions[64] that occur in propositions (categories of meaning), the ways in which complex notions are composed of simpler notions, the ways in which propositions are composed of notions, and the ways in which compound propositions are built up from simpler ones (the meanings of connectives, in current logical terminology). Now, since knowledge about a field of experience is possible only by way of a theory that describes the behavior of the objects of that field, their relations and their connections etc., it is clear that the "ideal constituents of a theory" (notions, propositions, truths, inferences, logical laws) will be "conditions of the possibility of knowledge in general." [65]

According to a *structuring* that is de facto quite close to that in contemporary logic textbooks, formal logic is conceived by Husserl as hierarchically articulated in three different *logical* levels.[66] Each of these levels is considered from two points of view, terminologically distinguished in his later *Formal and Transcendental Logic* as 'apophantic' (concerned with the forms of propositions, of proofs, and of theories as systems of propositions)[67] and 'ontological' (concerned with objects in general, with sets and relations among sets, with any kind of manifold, with any kind of connection, order, magnitude, ...).[68]

At the basis of this structuring we find a 'logical morphology' that, from the *apophantic* point of view, aims at determining (i) the 'pure categories of meaning', that is the fundamental semantic categories, (ii) the forms of connection between the 'lower elements of meaning' (i.e. names and predicates) in simple propositions, and (iii) the elementary forms of generating propositions from

---

[61]*PR* 236–237; *PRe* 232.

[62]*PR* 241; *PRe* 235.

[63]*PR* 240; *PRe* 234.

[64]Concerning our choice to refer to non-propositional components of propositions as notions see below Appendix 4, footnote 96.

[65]*PR* 239; *PRe* 234.

[66]Cp. Casari 1999; Ortiz Hill 2002, 87–88; Tieszen 2004, 26–34. As Tieszen rightly stresses, in *FTL* the third level of Husserl's stratification of 'objective formal logic' is constituted by what Husserl calls 'truth logic' (*Wahrheitslogik*), an attempt to identify *material conditions of truths* for judgements that are already established to be consistent. We will not consider this level here.

[67]Null & Simons 1982, 448.

[68]Loc. cit.

propositions (in particular those that are deductively relevant, i.e. by Husserl's lights, conjunction, disjunction and implication[69]). Thus, from the *apophantic* point of view, logical morphology deals with the possible forms of primitive propositions as well as with the purely formal and a priori laws that govern their possible compositions (the 'laws of complication'), *without raising the question of the truth or falsity of propositions at all.* From the *ontological* perspective, logical morphology aims at determining the pure or formal objectual categories, that is the higher genera 'object', 'state of affairs (*Sachverhalt*)', 'unity', 'plurality', 'number', 'relation', 'connection' etc., under which all conceivable objects and all conceivable 'states of affairs' must fall.[70]

On the second level, which is based upon this logical morphology, we find a discipline whose objects are laws that are to ensure, at the purely formal level – that is *without raising the question of the material truth or falsity of propositions* – the avoidance of contradiction ("[formal] truth or falsity of the meanings exclusively on the basis of their pure categorical form of construction"[71]). From the *apophantic* point of view, the discipline under consideration aims at characterizing *logical* laws as well as the laws that 'unify in a consistent way' several propositions in theories. From the *ontological* perspective, it has the task of establishing which mathematical structures are coherent from a point of view of logical admissibility (consistency) and which ones are not ("the being or not-being of objects in general, states of affairs in general etc., again, on the basis of their pure categorial form"). In *Formal and Transcendental Logic* this discipline is called 'logic of consequence' or 'logic of non-contradiction'. "These laws, which concern meanings and objects as such, with the widest universality conceivable, the universality of logical categories, are in themselves theories."[72] On this second level are located, on the side of meaning, the theories of logical inference, such as traditional syllogistics and the theory of propositional inferences,[73] and, on the side of the object, abstract mathematics, such as elementary number theory and set-theory ("the pure theory of pluralities which has its roots in the concepts of a plurality [*Vielheit*], the pure theory of numbers, which has its roots in the concept of a number – each of them by itself a rounded-off theory").[74] Husserl characterizes this level as follows:

---

[69]See *LV96*, 135–141. "The three forms of connection discussed above (sc. conjunction, disjunction and implication) are the only elementary ones for propositions in general" (140). Negation is considered as an operation, since it takes only one argument: "The operation of negation is applicable to any proposition, that is to every proposition corresponds its denial (negation). This is a proposition that has the original proposition as its topic (*Subjekt*) and denies its truth." Husserl also regards the affirmation of a proposition, that is the passage from 'S is p' to '[That] S is p is true', as an operation and writes: "Both affirmation and negation of a proposition ... [are] propositions about (*über*) propositions" (135).

[70]*PR* 244; *PRe* 237. Cp. also *FTL* [ed. 1929], 77–78.

[71]All quotations in this paragraph are from *PR* 245–246; *Pre* 237–239.

[72]*PR* 245–246; *PRe* 238–239.

[73]For a formal presentation of Husserl's theory of propositional inferences, see below Appendix 5.

[74]*PR* 246; *PRe* 239.

> All the laws that belong here [i.e. on second level, no matter whether viewed from the apophantic or from the ontological perspective] lead to a limited number of primitive or basic laws, which have their immediate roots in our categorial concepts. In virtue of their homogeneity, they must serve to justify an all-comprehensive theory, which will contain the particular theories just mentioned as relatively closed components.[75]

Levels one and two together correspond to (what in the terminology of current logic might be called) specification of a formal language and of a logical calculus, and with this, as Husserl says, "the idea of a science of the conditions of the possibility of a theory in general is dealt with sufficiently".[76]

Finally, on the third level, based in turn on the logic of non-contradiction, we find a 'theory of deductive theories' that has as its subject-matter, from the apophantic point of view, the a priori forms of possible theories and, from the ontological perspective, their objective correlates, that is the varieties or manifolds considered as the formal counterpart of a possible field of knowledge in general: "The objectual correlate of the concept of a possible theory, definite only with respect to its form, is the concept of a possible field of knowledge (*Erkenntnisgebiet*) controlled by a theory of this form."[77]

As regards the idea of an axiomatization of the theories that Husserl seems to have in mind, we can say that the formal theories appear to be conceived as deductive systems defined by a *finite number* of axioms. Manifolds, on the other hand, are conceived analogously to then current mathematical thought, as sets or as sets provided with some algebraic or topological structure[78], which underlie the theories intended as deductive systems. The last chapter of the *Prolegomena* provides strong evidence for the claim that Husserl has a clear idea of the distinction between a theory as formal system (i.e. a theory based on the concept of formal proof) and a theory as a collection of models (of some set of axioms), and that he conceives manifolds as classes of models corresponding to theories.[79] In general, a formal theory is determined by a set of formulae that constitute the axioms of the

---

[75] *PR* 246; *PRe* 239.

[76] *PR* 247; *PRe* 239.

[77] *PR* 248; *PRe* 241. Cp. also Ortiz Hill 2002, 88.

[78] Cp. Casari 2000. Ortiz Hill 2000a rightly stresses Husserl's distinction between pure sets in Cantor's sense and manifolds: "[Husserl] had come to clearly distinguish his manifolds from Cantor's *Mannigfaltigkeiten* or sets ..." (173), and in 2002 she writes: "Husserl's manifolds are not aggregates of elements without relations. It is precisely the relations that are essential and serve to distinguish a manifold from a mere aggregate.... Husserl saw manifolds as aggregates of elements that are not just combined into a whole, but are continuously interdependent and ordered ..." (97). Apparently Ortiz Hill, too, assumes that manifolds are sets provided with some topological structure. Husserl clearly marks the relevant distinction when he says in *FTL*: "From [a] particular field of objects [we obtain] the *form of a field* or, as the mathematician says, a manifold. It is not a mere manifold, *for that would be the same of a mere set ... Rather it is a set whose special feature is ... that it is conceived as 'a' field which is determined by* [a] *complete group ... of axioms-forms...*" (81).

[79] Cp. Null & Simons 1982, where an interpretation of manifolds as certain well-defined classes of relational structures is developed.

theory, on the one hand, and by the classes of its models, on the other. Husserl thinks that it is always possible not only to axiomatize a theory, but also to formulate and prove general theorems about the relations that obtain between various deductive systems. Indeed, he introduces purely formal relations among theories, respectively among manifolds, such as those of "generalization" and "specialization"[80] and, finally, explicitly poses as a task for the "theory of theories" to find relevant correspondences between certain abstract properties of formal theories and certain abstract properties of manifolds:

> These various forms (of theories) are not ... without mutual relations. There will be a determinate order of proceeding according to which we can construct the possible forms, survey their lawful interconnections and hence also move from one to another, varying certain fundamental determining factors, etc. There will also be ... general propositions that, for certain forms of theory, govern the ... connection and transformation of those forms... This is a last, highest goal for a theoretical science of theories in general.[81]

2.6 Logical Morphology and Logic of Non-Contradiction in the *Fourth Investigation*

From the 'apophantic' point of view, logical morphology has the task of determining the fundamental semantical categories, the forms of composition of simple meanings in complex meanings, including propositions, and, finally, all possible forms of elementary connections for propositions. So its main task is to formulate laws that regulate the construction of grammatically well-formed sentences *without* deciding whether their meanings are coherent – or rather contradictory, absurd or ridiculous. Among the questions to be answered are the following: How must a subject and a predicate be connected in order to form a grammatically correct sentence? In what manner is a meaningful complex term generated from simple terms? How is a meaningful compound sentence generated from simple sentences?

The starting point in the *fourth Investigation* is given by the distinction, known from Scholasticism, between categorematic terms that have a meaning by themselves, and syncategorematic terms that are applied to categorematic terms to produce new meaningful terms. It is important to stress that in Husserl's theory of meaning syncategorematic terms by no means lack meaning altogether.[82] Rather, they express *moments of dependent meaning*:[83] their meaning demands completion

[80]For a precise account of these notions see next chapter.

[81]*PR* 247; *PRe* 240.

[82]For Bolzano, too, syncategorematic expressions are meaningful, but Bolzano does not distinguish between dependent and independent meanings.

[83]This is an application to the field of meaning of the notion of dependence that was explained in the *third Investigation* with respect to objects in general. Dependent are "contents not able to exist alone, but only as parts of more comprehensive wholes" (*LU* IV, §7, 311; *LI* 506).

which it receives within a more complex expression that has an independent meaning.

In §10 of the *fourth Investigation* ("*A priori laws governing combinations of meanings*") Husserl develops the idea of abstracting the logical form from the observable grammatical structure of sentences. "To consider an example. The expression 'this tree is green' has unified meaning. If we ... proceed to the corresponding pure form of meaning, we obtain 'this *S is P*', an ideal form whose extension (*Umfang*) consists solely of independent [sc. sentential] meanings."[84]

This concept of a "sentential form (*Satzform*)" – in this case the result of substituting individual variables for the individual constants in a sentence and predicates variables for its predicate constants – captures a structure of complex meanings that does not vary when the components of these meanings vary. Any sentence whatsoever can thus be transformed into a sentence schema that subsumes all sentences of the same form and that represents what the propositions they express have in common.

Husserl's doctrine of semantic categories adopts a central idea of Bolzano's logic of variation, and it fills up a lacuna in his theory. Bolzano had shown that it is logically illuminating to consider the results of systematically varying some non-propositional components of a proposition within certain limits, that is within a homogeneous sphere of elements somehow specified in advance to which the component to be varied belongs. In a first approximation one can say that his logic of variation consists in the examination of the semiotic relationships that obtain between a proposition X and its variants when some components of X are replaced by others belonging to the same sphere of variation. But Bolzano himself did not precisely define what belonging to the same sphere amounts to.

For Husserl, a semantic category is constituted by the class of all expressions that can be substituted for a component of a meaningful sentence *salva congruitate*, that is without detriment to the grammaticality or meaningfulness of that sentence. Within the sentential schema used above, "we cannot substitute any meanings we like for the variables 'S' and 'P'... Any nominal material (*Materie*) ... can here be inserted (sc. for 'S'), and so plainly can any adjectival material replace the 'P' ... but if we depart from the categories of our meaning-material, the unitary sense vanishes. ... In such free exchange of materials within each category, false, foolish, ridiculous meanings ... may result, but such results will necessarily be unified meanings."[85] By contrast, the string of words 'but or similar and' lacks a coherent or "unified" meaning (as Husserl puts it), even though the words it consists of are meaningful.[86]

Husserl's logical morphology clearly anticipates the concept of a formal language as well as the modalities of its constitution, and it investigates some of its

---

[84]*LU* IV, §10, 318–319; *LI* 511.

[85]*LU* IV, §10, 319; *LI* 511–512. Cp. Tieszen 2004, 26 ff.

[86]The categorization of linguistic expressions that is invoked here could be usefully compared to the typification of entities in Russell's theory of types.

main features. In general, a formal language is constituted by the specification of its alphabet and its well-formed expressions, and it performs two fundamental functions, the indicative one (indication of individuals, properties and relations) and the declarative one (declaring that things are in a certain way). The constitution of formal languages then goes through the following process: a certain initial set of words is specified, and some generation procedures are specified through which it is possible to obtain new words, requiring that only the words in the initial set and those obtainable through iteration of the specified procedures must be admitted in the formal language. The cumulative nature of this kind of constitution is obvious: under the assumption of closure with respect to the iterable operations, the definition procedure just described is capable of generating all possible constructs, starting from "fundamental forms".

The definition of primitive terms in a formal language corresponds *grosso modo* to Husserl's way of specifying the primitive kinds of meaning categories. The meaning categories that matter to Husserl largely coincide with the classes of expressions that current logic deems to be indispensable, namely subjects (singular terms) and predicates on the categorematic side and connectives and the other logical operators on the syncategorematic side. What Husserl refers to as "*a priori* laws ... that govern the combination of meanings into new meanings,"[87] can be considered as the counterpart to the procedures that allow the generation of new expressions in a formal language.

The laws at the first level of logic that are to inhibit the formation of terms or sentences that are grammatically ill-formed (*Unsinn*) constitute the "ideal scaffold (*ideales Gerüst*)"[88] of language, the ideal structure of various actually existing natural languages. Husserls regards this as an attempt to execute the programme of a "universal grammar" that was developed by rationalists in the seventeenth and eighteenth century.[89]

The second level of formal logic, the logic of consequence or of non-contradiction, as it is called in *Formal and Transcendental Logic*, is constituted on the apophantic side (the only one we will be treating here) by laws that are to ensure the avoidance of contradictions (*Widersinn*). Since Husserl uses Bolzano's term 'objectual' (*gegenständlich*)' for stating that a term denotes something,[90] we can say that at this level formal logic is concerned to establish under which conditions a meaningful complex term can be objectual. It also seeks to answer the question under which conditions a meaningful sentence can express a truth, – in other words, it seeks a criterion to decide whether a proposition is formally true or formally false. Laws at this level determine whether any object can correspond to the linguistic constructs built up in conformity with the laws established at the first level. It is

---

[87] *LU* IV, §10, 317; *LI* 510.

[88] *LU* IV, §14, 338; *LI* 526.

[89] *LU* IV, §14, 336; *LI* 524.

[90] For Bolzano objectuality is primarily a property of notions (*Vorstellungen an sich*): a notion is objectual if and only if there is an object that falls under it, and objectless otherwise.

clear that if the constructs unite incompatible constitutive elements, then no object can correspond to them; but if this does not happen, if, as Husserl puts it, "the meaning has *objective* validity (*objective Gültigkeit*),"[91] then the construct *can* have an object. Whether such an object de facto exists is a problem that is not decided at this level.

Husserl strictly distinguishes incompatibility (*Unverträglichkeit*) as nonsense (*Unsinn*), which results when the formation-rules for complex terms and sentences are violated, from incompatibility as contradiction or "countersense (*Widersinn*),"[92] and he divides the latter into two kinds: what suffers from *Widersinn* is either a formal or analytic contradiction ("formal countersense"), or it is a material or synthetic contradiction ("material countersense").

Husserl agrees with Bolzano that Kant's attempts at explaining 'analytic' "do not deserve to be called 'classical'".[93] He distinguishes "analytic laws", which are pure analytic truths, from "analytic necessities", which are impure analytic truths. The purity of the former consists in the fact that they contain only formal concepts. Impure analytic truths are said to be "formalizable *salva veritate*", that is they can be transformed into pure analytic truths by replacing the material concepts they contain by formal concepts. Thus, for example, the proposition that (if Socrates has both courage and wisdom then he has courage) is an impure analytic truth, an "analytic necessity". It can be formalized *salva veritate*, the upshot of this procedure being the pure analytic truth, the "analytic law", that (for any properties *x* and *y*, if something has both *x* and *y* then it has *x*). So what are generally referred to as analytic propositions are "particularizations (*Besonderungen*)" of analytic laws. Husserl's analytic necessities are expressed by the logically valid sentences of contemporary logic: what such sentences express is true, no matter what their descriptive parts actually mean, – it is true, as the saying has it, "in virtue of logic."

By contrast, an analytic contradiction is a falsehood that contains no material concepts (it is false "in virtue of logic"), or it is formalizable *salva falsitate*. Thus the proposition that (if something has a certain property then it lacks it) and the proposition that (if Socrates has courage then he lacks courage) are examples of analytic *Widersinn*.

For Husserl, a synthetic law a priori is a "law that contains material concepts in such a way that does not allow their formalization *salva veritate*". Particularizations of such laws are synthetic necessities. So the truth that (if something is clearly red all over then it is not clearly green all over) is a synthetic law a priori, and the truth

[91] *LU*, IV, 294; *LI*, 493.

[92] "One must, of course, distinguish the ... incompatibilities to which the study of *syncategorematica* has introduced us, from the other incompatibilities illustrated by the example 'a round square'" (*LU* IV, §12, 326; *LI* 516).

[93] All quotations in this paragraph and the next two are from *LU*, 3, §12 ('Basic determinations concerning analytic and synthetic propositions'), 254–256; *LI* II, 457. On Bolzano's account of analyticity see Morscher 2008, 60–63, 161–167, Künne 2008, 233–304 and, for a comparison with Husserl, Künne 2009, §§3–4.

that (if this patch here is clearly red all over then it is not clearly green all over) is a synthetic necessity.

By contrast, the proposition that (if something is clearly red all over then it is clearly green all over) and the proposition that (if this patch here is clearly red all over then it is clearly green all over) are examples of material *Widersinn*. The sentences expressing these propositions are allowed by Husserl's logical morphology: they are grammatically impeccable. The same holds *mutatis mutandis* of the term 'round square': it cannot possibly apply to anything, and yet it is grammatically flawless, hence permitted by logical morphology.

About the relation between the logical order and the real order Husserl writes: "The consistency or absurdity of meanings implies objective and a priori possibility (consistency, compatibility) or objective impossibility (incompatibility); in other words, it implies the possibility or impossibility of there being objects that are meant ..., to the extent that this depends on the intrinsic essence of those meanings."[94]

2.7 Appendix 4: On Bolzano

2.7.1 The Relation of Derivability (Ableitbarkeit)

For Bolzano the most important relations between *propositions* (*Sätze an sich*) come to light when one considers certain notions they contain as variable, that is as replaceable by others, and then asks in which relation the propositions obtained by variation stand to truth or falsity.[95] One of the earliest uses of the operation of systematic variation (*Veränderung*) in Bolzano's *Wissenschaftslehre* is to be found in his theory of "ideas in themselves (*Vorstellungen an sich*)", which we call *notions* for the sake of brevity:[96] here the aim is that of extending concepts originally introduced only for those notions that are non-empty or objectual (*gegenständlich*) to those which are empty or objectless (*gegenstandlos* [sic]).[97] One of the most significant applications of this operation, however, is meant to bring to light certain logically important relations among propositions.

---

[94] *LU* VI, §14, 334; *LI* 523.

[95] *WL* II, §154, 100.

[96] Notions are either (objective) concepts or (objective) intuitions in themselves (*Anschauungen*). Intuitions are said to be notions that are simple and have exactly one object, and concepts are defined as notions that are not intuitions and do not contain any intuition as part. An intuition is expressed in an utterance of 'this' if the demonstrative is used to refer to something perceptually given. Cp. *WL* I, §§72–78, 325–360.

[97] *WL* I, §108, 513–515. It was invoked for the first time in *WL* I, §66, 299–300 where the topic of indexicality is briefly touched: the notion that is now expressed by "a presently living human" is replaced by another notion when this phrase is uttered at a different time, since the time-specifying component of the former notion is varied.

In §155 of his *Wissenschaftslehre* Bolzano introduces the relation of derivability (*Ableitbarkeit*)[98] as a special case of the relation of compatibility (*Verträglichkeit*). As a result of this decision, in Bolzano's logic nothing is derivable from incompatible premises. A collection of propositions is said to be compatible if and only if there is *at least one* substitution of some or all their extra-logical parts[99] that makes all of them simultaneously true. (Thus the propositions expressed by 'Socrates is taller than Phaedo' and 'Phaedo is taller than Socrates' are compatible in the Bolzanian sense, since substituting the notion expressed by 'knows' for the notion expressed by 'is taller than' results in two truths).

The relation of derivability is introduced as follows:

> Let us consider ... the case that among the compatible propositions A, B, C, D, ..., M, N, O, ... the following relation obtains: all notions whose substitution for the variable notions i, j, ... turns a certain part of these propositions, namely A, B, C, D, ... into truths, also have the property of making a certain other part of these propositions, namely M, N, O ... true. ... To this special relation ... I wish to give the name of *derivability* (*Ableitbarkeit*)... Hence I say that propositions M, N, O are *derivable* from propositions A, B, C, D, ... with respect to the variable parts i, j, ..., if every collection (*Inbegriff*) of notions whose substitution for i, j, ... makes all of A, B, C, D, ... true, also makes all of M, N, O, ... true.[100]

In other words, the relation of derivability obtains between two (sets of) propositions with respect to a certain series of notions if and only if (i) the premises are compatible with respect to that series of notions[101] and (ii) every series of notions that makes all the premises true when substituted for the chosen series also makes all the conclusions true.

Bolzano proves a great number of theorems concerning the formal relation of derivability. Here are two examples. There is a generalized rule of transitivity for "Bolzanian sequents", which we can conceive as syntactical objects of the form $\Gamma^\wedge \rightarrow \Delta^\wedge$ (i.e. the conjunction of the formulae in the antecedent implies the conjunction of the formulae in the consequent):

---

[98]Nowadays this terminology is prone to cause a misunderstanding, since it has become customary to use this term in a purely syntactical sense. 'Deducibility' would have a strong syntactical connotation, too; whereas for Bolzano derivability is a (quasi-)semantical relation. Nevertheless, we have preferred to stay close to Bolzano's wording.

[99]According to Bolzano all propositions can be expressed by instances of the schema 'A has b' where 'A' expresses any notion (of whatever complexity), while 'b' expresses a notion of a property (*Beschaffenheit*). Bolzano's copula 'has' expresses the notion of exemplification which is a logical notion. In *WL* II, §148, 84 Bolzano maintains that there is no sharp line of demarcation between logical and extra-logical notions, but he leaves no doubt that his copula expresses a logical notion.

[100]*WL* II, §155, 113–114.

[101]In the quoted passage, Bolzano explicitly requires compatibility also for the consequences M, N, O, ... This, however, already follows from the definition of derivability. We say that certain propositions follow from certain others when (i) there is a substitution that makes the premises true (compatibility); (ii) all substitutions that make the premises true, also make the conclusions true. From (i) and (ii) follows: (iii) there is a substitution that makes both the premises as well as the conclusions true.

$$\frac{\Gamma^\wedge \rightarrow \Delta^\wedge \quad \Delta^\wedge \wedge \Theta^\wedge \rightarrow \Phi^\wedge}{\Gamma^\wedge \wedge \Theta^\wedge \rightarrow \Phi^\wedge}$$

And a more sophisticated version:
If the notions i, j .../ m, n .../ p, q ... are all mutually different, the following theorem holds:

$$\text{if } \Gamma \vDash_{i,j,\ldots m,n} \Delta \text{ and } \Delta + \Theta \vDash_{m,n,\ldots p,q,} \Phi \text{ then } \Gamma + \Theta \vDash_{i,j,\ldots m,n,\ldots p,q} \Phi$$

on condition that p, q, ... do not occur in Γ and i, j ... do not occur in $\Delta + \Theta$. The simplest case seen above is the one where i, j, ... and p, q, ... are empty.[102]

It has been remarked repeatedly in recent studies,[103] that this is a first characterization of the notion of logical consequence, which differs from the one given by Tarski in various respects. (i) Bolzano's explanation of the notion of "following from certain premises through a correct inference" rests on the concept of 'variation' of notions, whereas the model-theoretic version of Tarski's definition of truth is concerned with non-interpreted languages and so, in order to define truth, or rather satisfaction, one has to put that language in correspondence with an "external" (set-theoretic) structure. (ii) Bolzano also allows for varying *only some* extra-logical notions contained in a proposition: the concept of "following from certain premises" is then relativized to a certain set of notions, while the basically Tarskian account of the concept of logical consequence given in current logic corresponds to a special case of the Bolzanian account, namely that in which *all* (and only) extra-logical notions are varied.

By admitting derivability not only with respect to *all* the extra-logical components (formal or logical derivability) but also with respect to just *some* of them Bolzano can cope with *material* or *enthymematic* inferences.[104] Thus, that Socrates is mortal is derivable from the premise that Socrates is a man, for every replacement of the notion expressed by 'Socrates' that makes the premise true also makes the conclusion true. This is not a case of logical derivability but a case of enthymematic derivability. Of course, it can be transformed into a fairly well-known case of logical derivability by adding the premise that all men are mortal.

In §60 of the *Logikvorlesung* of 1896 (*Logical and illogical inferences and corresponding divisions of hypothetical truths*) Husserl tries to characterize the difference between formal/logical derivability and semiformal/material derivability (relative to certain notions). At first we find a characterization of logically valid inferences:

> The hypothetical proposition "if every man is mortal and Socrates is a man, then Socrates is mortal" is a truth. But here we have the peculiarity that in this truth certain moments of the matter (*Momente der Materie*), namely the presentations or concepts Socrates, Man and

---

[102] *WL* II, §155, 122–123.

[103] Berg, in: BGA I, 12/1, 26; Casari 1985; Paoli 1991; Siebel 1996.

[104] Cp. George 1983.

> Mortal can be varied at will. . . . The particular propositions which constitute the matter of the hypothetical truths can be false, absurd or ridiculous, nonetheless it always remains a hypothetical truth. At its foundation there is indeed a law that we express in the form: "If all A are B and S is an A then S is a B". . . . [T]he form of the connection is of a general logical kind, i.e. it has a meaning that is independent from the specifics of any class of matter (*Klasse von Materien*). This is not always the case.[105]

And then Husserl goes on to characterize enthymematic derivability:

> if I compare three men, let's say Hans, Kunz and Wilhelm, and I infer: Hans is taller than Kunz, who is taller than Wilhelm, hence Hans is taller than Wilhelm, this inference is, likewise, a particularization of a law. . . . However, a relation of magnitude is meaningful only for something that has magnitude. More exactly: such a relation applied to things that do not have any magnitude leads to absurd representations. Hence the relation of magnitude depends on the specifics of the matter (*Besonderheit der Materien*). The inference "a > b, b > c, therefore a > c" does not apply to the general logical domain, it is limited to the field of magnitudes.[106]

If the language is interpreted and '>' expresses the notion *taller than*, the inference "$a > b$. $b > c$. Therefore $a > c$" is not logically but enthymematically valid. That it is enthymematically valid becomes clear if we take the notions *a, b, c* (i.e. only *some* of the extra-logical notions involved) as variable and keep the notion *taller-than* fixed. It is clear as well that the inference can be transformed into a logically valid one by adding as an additional premise a claim as to the transitivity of the relation ascribed by ">". By contrast, the argument is not logically valid, for if *all* its extra-logical components were variable *salva veritate*, then all relations would have to be transitive.

2.7.2 The Relation of Exact Derivability (genaue Ableitbarkeit)

In §155 no. 26 of his *Wissenschaftslehre* Bolzano considers a strengthening of the relation of derivability and thereby confronts and solves the problem of *redundancy*:

> Let proposition M be derivable from premises A, B, C, D, ... with respect to the notions i, j, . . . If A, B, C, D, . . . are such that none of them, nor even any of their parts, may be omitted, with M still being derivable form the remainder with respect to the same notions i, j, . . . , then I call the relation of derivability of proposition M from A, B, C, D, . . . exact (*genau*), irredundant (*genau bemessen*) or adequate. In the opposite case I call the relation *redundant* (*überfüllt*).[107]

Exact derivability holds between a set X of propositions and a *single* proposition p with respect to a certain list $\underline{a}$ of notions if and only: (i) p is *derivable* from X with respect to $\underline{a}$ and, moreover, (ii) for no *proper subset* Z of X it holds that p is derivable from Z with respect to $\underline{a}$; and (iii) for no *proper sublist* $\underline{b}$ of $\underline{a}$ it holds that

[105]*LV96*, 238–239.

[106]*LV96*, 239.

[107]*WL* II, §155, 123.

p is derivable from X with respect $\underline{b}$. So the conclusion cannot be derived either from a proper subset of the premises or with respect to fewer notions. In other words, all premises should be relevant *in toto*.

In the previous paragraph we assimilated the Bolzanian notion of derivability to a (generalization of) the Tarskian notion of logical consequence. Now while Tarski's notion of logical consequence satisfies *monotonicity*:

$$\Gamma \vDash p \text{ implies } \Gamma + \Delta \vDash p$$

the Bolzanian notion of exact derivability does not:

$$\Gamma \vDash^{genau} p \text{ does not imply } \Gamma + \Delta \vDash^{genau} p.$$

While general derivability occurs between sets of propositions, Bolzano takes the consequent of a relation of exact derivability always to be a single proposition. This request is intended to comport with mathematical practice: in a rigorous mathematical proof what is derived is always *one* proposition. Bolzano proves a number of theorems concerning exact derivability. Let us consider here only the most important ones[108]:

(i) Neither the premises nor the conclusion of an inference of the kind *genaue Ableitbarkeit* can be a logical truth, for if a logical truth were among the premises it would be superfluous and if the conclusion were a logical truth then all premises would be superfluous.
(ii) The premises of an inference of the kind *genaue Ableitbarkeit* should be reciprocally independent, that is none can be derived from the others.
(iii) The rule of transitivity (hence also the generalized transitivity rule for Bolzanian sequents considered in the previous section) doesn't hold in general for exact derivability, as Bolzano shows in §155 no. 32 by exhibiting a suitable counterexample:

> If the relation of derivability between premises A, B, C, D, ... and the conclusion M, and also the relation between premises M, R, S, T, ... and the conclusion X is exact with respect to the same notions i, j, ..., it does not follow that the relation of derivability, which holds between premises A, B, C, D, ... R, S, T, ... and the conclusion X,[109] must also be exact. For example, the relation of derivability between the premises: all α are β, all β are γ; and the conclusion: all α are γ; and also that between the premises: all α are γ, all γ are β and the conclusion: all α are β, are without doubt exact. But the relation of derivability between the three premises: all α are β, all β are γ, all γ are β and the conclusion: all α are β, is not exact.[110]

That is, there are suitable Γ, Δ and p, q such that, in symbols:

$$\Gamma \vDash^{genau} p \text{ and } \mathrm{p} + \Delta \vDash^{genau} q, \text{but } \Gamma + \Delta \nvDash^{genau} q.$$

---

108 Cp. Paoli 1991, 233–234.

109 Compare the proof given above.

110 *WL* II, 125–126.

Bolzano's notion of exact derivability (*genaue Ableitbarkeit*) provides for an understanding of "if... then" that is closer to the notion we have before we learn classical logic. In this sense, Bolzano shares with Relevance Logic the concern of finding a concept of implication capable of avoiding the paradoxes of material implication, and in this respect he comes closer than Frege to our pre-logical intuitions. From the standpoint of current logic, the Fregean conditional (*hypothetisches Satzgefüge*) seems to be nothing but Russell's truth-functional, material implication. Actually, this is not quite right, since the Frege of *Grundgesetze* conceives of the main operator in a conditional not as a *connective* in the currently standard sense (i.e. an operator producing a sentence out of any two given sentences), but as a binary *functor* defined on the totality of objects which, applied to the objects denoted by the sentences A and B, in this order, gives as value the False, in case A denotes the True and B denotes anything other than the True, and in all other cases gives as value the True. But whether one takes sentences to be a special kind of proper names or not, the Fregean conditional "the sun shines $\rightarrow 3 \times 7 = 21$" expresses a true proposition, in spite of there not being any relevant connection between the contents of the antecedent and the consequent.

2.7.3 *The Relation of Consecutivity* (Abfolge)

Bolzano introduces the relation of consecutivity (variously called "grounding", "ground-consequence" or "objective consequence") in *Wissenschaftslehre* II, §162, and he develops it, albeit incompletely, in the same volume in §§198–222. Following Casari, this doctrine may be called Bolzano's aetiology.[111]

"There is a noteworthy relation that obtains among truths ..., in virtue of which some truths are related to others as *reasons* [or *grounds*] to their *consequences*".[112] This relation does not only obtain between two truths but also among collections of truths, in which case the members of the two collections are partial reasons/consequences (*Theilgründe*, *Theilfolgen*) and the collections are the complete reason/consequence (*vollständiger Grund*, *vollständige Folge*).

The relation of consecutivity can be expressed by 'because' sentences: "we say that A is because B is, when we want to say that the complete or a partial reason of the truth A resides in the truth B".[113] Actually, we hardly ever use 'because' sentences instantiating Bolzano's schema in which the letters are placeholders for *names* of propositions rather than expressions of propositions, that is sentences. What could 'A is because B is' then mean if not 'A is *true* because B is *true*'? But Bolzano's examples are not of this form: they instantiate the simpler schema '*p*,

[111]For illuminating accounts of this doctrine see Buhl 1961; Berg 1962, 151–164; Casari 1992, Sebestik 1992, Pt. 2, Chapter 4, Mancosu 1991; Tatzel 2002.

[112]*WL* II, §162, 191.

[113]*WL* II, §177, 221–222; cp. §168, 207.

because *q*'. So let us modify his contention slightly: We say that (*p* because *q*), when we want to say that the complete or a partial reason of the truth that *p* resides in the truth that *q*.

The relation of consecutivity is a relation that obtains exclusively among truths. By contrast, the relation of derivability can also obtain between false propositions. Let us consider Bolzano's favourite example for making the difference between consecutivity and derivability salient. "[T]he truth that in summer it is warmer than in winter contains the ground of that other truth that the thermometer stands higher in summer than in winter, while the latter truth can be considered as a consequence of the former".[114] Of course, if one knows that the thermometer stands higher in summer than in winter, one can infer/derive the truth that in summer it is warmer than in winter, but nobody would acknowledge the first of these truths as the, or a, reason for the second. By recognizing (*erkennen*) that the 'thermometer' proposition is true we could acquire the knowledge *that* the 'warmth' proposition is true. But the 'thermometer' proposition does not explain *why* the 'warmth' proposition is true. "A truth which is related to certain other truths as a consequence to its reasons is frequently also *derivable* from these latter truths, provided that we envisage certain notions as variable. The proposition that the thermometer stands higher in the summer than in winter is obviously derivable from the proposition that in summer it is warmer than in winter, provided that only the notions of summer and of winter are considered variable".[115] But in this case the relation of derivability also holds in the other direction: these two propositions are interderivable with respect to those extra-logical notions. By contrast, the relation of consecutivity is *asymmetric*.[116]

Bolzano sometimes uses another example that is in itself philosophically interesting. "Let A be any truth you like: the truth that the proposition A is true is a proper consequence of A."[117] So according to Bolzano, it is true that *p*, because *p*, – in other words, that things are thus-and-so explains why it is true that things are thus-and-so. Again, the propositions in such pairs are interderivable with respect to all extra-logical notions contained in the proposition that *p*. But the ground-consequence relation, here as always, obtains only in one direction. Furthermore, unlike derivability, consecutivity is also *irreflexive*: nothing is its own ground.[118] If Bolzano is right about such pairs, Frege is wrong, for he famously maintained that the thought that it is true that *p* is identical with the thought that *p*.[119]

Let us register some other important features of consecutivity that Bolzano points out: This relation is *intransitive* ("one cannot say of a consequence of

[114] *WL* II, §162, 192.

[115] Loc. cit.

[116] *WL* II, §203, 352; §209, 362.

[117] *WL* II, §205, 357; cp. §212, 370, §214, 374. More on this topic in Künne 2003, 46, 151–152.

[118] *WL* II, §204, 356.

[119] Cp., for example, Frege, *Der Gedanke* (1918), 61 (original pagination).

a consequence [...] that it is the consequence of the *reason* of its reason, without altering the concept"[120]). It satisfies the conditions of

- *Functionality*: every truth is a reason of at most one other truth[121]
- *Injectivity*: every truth has at most one reason[122]
- *Non-surjectivity*: there are truths that do not have a reason, the so-called fundamental or primitive truths (*Grundwahrheiten*), of which there is more than one, Bolzano thinks, though he admits of having no proof for this,[123] and
- *Non-monotonicity*: "we must not regard a consequence that follows from several truths A, B, C, ... as a consequence of the whole collection of truths A, B, C, D, E, F, ..."[124]

Is the relation of consecutivity a special case of the relation of derivability? Bolzano takes this to be probable, but he admits that he knows no proof for it.[125] If the affirmative answer were right then it would not only be the case that all substitutions of certain notions that make the premises true render the conclusions true as well, but also that these substitutions always result in truths that are related as reasons to consequences. But is it really "probable", Bolzano wonders, "that for every collection of truths from which another truth follows as from its reasons, there are infinitely many other collections of truths from which other truths follow in one and the same way, namely in such a way that the peculiarities (*Besonderheiten*) of the notions which these collections of truths consist of never have any influence upon the way they are related as ground and consequence?"[126] In other words: consecutivity is a relation that in some way depends on the content of the connected truths, on "the peculiarities of the [sc. extra-logical] notions" which are the components of those truths, whereas the relation of derivability can account for a conception of "following from certain premises through correct inferences" as a relation on which such peculiarities have no influence.

The relation of consecutivity is not to be identified with that of causality.[127] The concepts of cause and effect apply only to objects that have "existence (*Dasein*)", which Bolzano identifies with "actuality (*Wirklichkeit*)", that is the ability of acting upon something (*Wirksamkeit*). By contrast, the relation of consecutivity obtains only between causally impotent entities, namely (true) propositions. However, if two actual objects stand in the relation of cause-effect, then the truths that ascribe

[120]*WL* II, §213, 371.

[121]*WL* II, §206, 359.

[122]*WL* II, §206, 359–360, though it can happen that different reasons share some partial consequences.

[123]*WL* II, §214, 374–376.

[124]*WL* II, §207, 360.

[125]See also the incipit of §201, 349: "If the relation of consecutivity is not a species of the relation of derivability, one cannot hope to explain the former in terms of the latter; hence one must look for other cognate concepts".

[126]*WL* II, §200, 348.

[127]*WL* II, §168, 208; §201, 349–350.

existence to those objects stand in the relation of consecutivity. For example if on the level of actuality God is the *cause* of the world, then on the level of the propositions the truth that God exists is the *ground* of the truth that the world exists. In a way Bolzano is trying to transfer the peculiar nexus intrinsic to the concept of causality to the level of a logical relation of dependency among true propositions.[128]

As in so many other respects, Bolzano stands here on Leibniz's shoulders:

> A reason is a known truth whose connection with some less well-known truth leads us to give our assent to the latter. But it is called a 'reason', especially and par excellence, if it is the cause not only of our judgement but of the truth itself... *A cause in the realm of things corresponds to a reason in the realm of truths*, which is why causes themselves ... are often called 'reasons'.[129]

Bolzano does not provide an analytic definition (*Erklärung*) of the concept of consecutivity, which would have to identify the components the concept contains and their mode of composition. He takes it to be probable, though, that the concept of *Grund* is simpler than the concept of *Folge* ("for generally one is not tempted to explain the ground through the consequence, but rather the consequence through the ground"[130]), and he reckons with the possibility that this concept is simple. Rather than defining the concept of consecutivity, Bolzano presents a bunch of postulates that constitute as many rules for its use. In these postulates he identifies a number of properties of consecutivity, such as being asymmetric, irreflexive, etc., many of which we have mentioned in this section.

2.7.4 Some Remarks on the Structure of Etiological Proofs

Etiological proofs are conceived as procedures to finding the grounds or reasons of a truth. Bolzano explicitly admits that sometimes such a procedure halts when reaching the so-called fundamental truths or principles. He also allows for processes of backtracking from a truth to its reasons that go on indefinitely. When represented graphically, etiological proofs have a tree structure that is very similar to that of proofs in a Gentzen-style sequent calculus.[131] In the procedure of tracing back the dependencies of a truth to its reasons, *immediate reasons* of a certain truth are those upon which it depends directly, and *auxiliary truths* (*Hülfswahrheiten*) are those on which its reasons depend. In one and the same proof a certain auxiliary truth (the same premise) can occur more than once (it is a partial reason of different

[128]Cp. Casari 1987, 332.

[129]Leibniz 1704, Book IV, Chapter xvii, §3.

[130]*WL* II, §202, 351.

[131]Introduced in G. Gentzen, *Untersuchungen über das logische Schließen*, 1935. Comparing this with WL II, §220, 380–383, one should keep in mind that Gentzen's notion of a "normal proof" is a *syntactical* concept.

consequences), but no truth can be an auxiliary truth of itself.[132] Hence Bolzano's logic is not, as we would say in current logical terminology, *resource-conscious*. In a proof we can use an assumption or hypothesis A as many times as necessary. The use of A does not, so to speak, *wear out* A.[133]

Bolzano uses the term "dependency" (*Abhängigkeit*) for the transitive version of the relation of consecutivity (*Abfolge*). The relation that obtains between truths in etiological proofs is the relation of *Abhängigkeit*, with respect to which Bolzano formulates, as surmises (*Vermuthungen*), some requirements or formal conditions. Especially important for a correct grasp of the concept "interconnection of truths" as explained in Chapter XI of the *Prolegomena* (*The Idea of Pure Logic*) are the following conditions:

(i) The requirement of non–increasing complexity (analyticity).

"I think that every purely *conceptual truth* (*Begriffswahrheit*) on which a second one depends, must never be *more complex* (*zusammengesetzer*) than the latter, though it need not be simpler. Propositions which constitute the objective grounds of a purely conceptual truth ... must not contain, each on its own, more parts than the truth that depends on them."[134]

(ii) The requirement that the reasons upon which a certain truth depends are not only the simplest, but also the most general.

"In a true and proper scientific exposition we must proceed from the more general to the more specific ... The more simple and more general truths are the foundation (*Grund*) for the more specific and more complex."[135]

Other requirements are (iii) compatibility of the premises (*reductio ad absurdum* proofs are not desirable)[136] and (iv) non-redundancy of proofs[137] (in a proof only premises that are necessary to reach the conclusion must be used, and only those principles that are deductively weakest). – Bolzano ends his discussion with the following remark:

> I occasionally doubt whether the concept of consecutivity, which I have above claimed to be simple, is not complex after all; it may turn out to be none other than the concept of an ordering of truths which allows us to deduce from the smallest number of simple premises the largest possible number of the remaining truths as conclusions.[138]

---

[132]*WL* II, §§217–219, 377–380.

[133]This problem is central for an important family of non-classical logics, so called substructural logics, which includes (among the others) linear logic and many-valued logics. Indeed, linear logic is both resource-conscious and attentive to the problem of relevance, while many-valued logics are (usually) resource-conscious but not attentive to the problem of relevance. Bolzano's logic is sensitive to the latter problem, but it is not, as we just saw, resource-conscious.

[134]*WL* II, §221, 384.

[135]*WL* II, §221, 388 note.

[136]*WL* IV, §530.

[137]*Beyträge* II §28, WL II, §223, 391–395.

[138]*WL* II, §221 note, 388.

2.8 Appendix 5. The Theory of Propositional and Conceptual Inferences in the *Logikvorlesung* of 1896

Although Husserl's *Logikvorlesung* of 1896 follows in many respects the model of Bolzano's *Wissenschaftslehre*, in the concluding section on the "doctrine of inferences (*Lehre von den Schlüssen*)" its base is the algebra of classes of Boole-Schröder rather than Bolzano's proof theory. In other words, the "deductive calculus" that Husserl outlines[139] is of an "axiomatic-synthetic" and not of an "analytical" kind.[140] Of this doctrine we want to examine:

(i) Husserl's remarks on the notion of *calculation* (*Rechnung*) as mechanical procedure and on the way this notion applies to derivations,
(ii) his theory of propositional inferences (*Theorie der propositionalen Schlüsse*) and,
(iii) his theories of predication and of conceptual inferences (*Theorie der konzeptualen Schlüsse*).

2.8.1 *The Concept of a Calculus*

How is a theory of inferences constituted? Husserl answers this question by using arithmetic as a model. In the same way as the analysis of the simplest arithmetical propositions shows that certain compositions (*Verknüpfungen*) ($+$, $-$, $\times$, : etc.) and relations ($=$, $\geq$, $\leq$)[141] are grounded a priori in the concept of number and obey certain laws (commutativity, associativity, etc.), the analysis of the structure of inferences shows that certain connections ($\wedge$, $\vee$ ...) and certain relations ($\rightarrow$)[142] are rooted a priori in the concept of proposition, and that their behavior is determined by certain laws. Hence one tries to isolate a minimal number of axioms capable of constituting the *principles* of the theory. The systematical progression from principles to derived laws occurs in one case through the mechanism of *calculation*, in the other through the mechanism of *deduction*, considered as the precise analogue of calculation.

Husserl also raises the question – and this is highly significant – whether these theories, of arithmetic and of inferences, are *adequate*. He observes that we lack a proof that the axioms of arithmetic are all those and just those conceivable in general for the characterization of this theory, that we lack a proof that the laws

---

[139] An analogous calculus is developed in the *Logikvorlesung* of 1902.

[140] This contrast in current logic corresponds to that between a Frege-Russell-Hilbert style calculus and a Gentzen style one.

[141] Cp. *PoA* 480; *PdA* 476.

[142] Here the symbol '$\rightarrow$' stands for the *relation* of *conditionality* (*Bedingtheit*), or 'inferability' between propositions. Cp. *LV96*, 254. See also Section 2.8.2.2 below.

which govern the behavior of the operations are the *only* laws which are valid for arithmetical operations, and that we lack a proof that the axioms of arithmetic are mutually *independent.*[143]

> Arithmeticians have gone through an infinite amount of trouble to establish the minimal number of arithmetical axioms. Yet it is still in doubt whether or not one of the axioms is simply a consequence of the others, without them having noticed it. At least, we still lack a systematical proof.

> The mathematician ... must not claim that what he offers as foundations is, for *a priori* reasons, everything that exists as foundation with respect to numbers *a priori* and in itself (*alles, was in Betreff der Zahlen and sich und* a priori *an Grundlegungen existiere*)... Exactly the same holds for the theory of propositional inferences and for everything else that we still have to discuss [i.e. for the theory of conceptual inferences]. We cannot guarantee completeness. ... And of course, as in arithmetic, there also is the task of shaping, as systematically as possible, the progress from basic laws (*Grundgesetze*) to derived ones and to develop methods that enable us to solve every conceivable problem (*Aufgabe*) with an orderly procedure and hence ... to prove deductively every inference we are presented with, no matter how complicated it is, i.e. to reduce it to elementary inferences.

The very same problems reappear in the constitution of the theory of propositional inferences.

As for the concept of a calculus Husserl recalls in his lectures some observations that we already met in his *Philosophy of Arithmetic* and that are also to be found in his 1891 review of Schröder's first volume of the *Vorlesungen über die Algebra der Logik.*[144] In both texts the core contention is that the deductive mechanism behaves exactly the same as the mechanism of calculation. In his review Husserl criticized Schröder for not having understood the nature of the *Folgerungskalkül*, in spite of having brilliantly improved its technical apparatus. It is not – Husserl objected – a "completely reformed deductive logic, in the form of an 'algebra of logic'", but rather an extrinsic, purely *mechanical* processing, a transformation of signs into signs according to certain rules, which can also be performed by a machine; it is a way of "sparing oneself (*sich ersparen*)" the trouble of making actual inferences, rather than a way of executing them. While "in [a machine] no thought corresponds to the signs", for *us* the result of the calculation expresses a proposition that enlarges our knowledge, as Frege would put it. In his *Logikvorlesung* Husserl says: "Calculating is operating with signs and not with the concepts themselves, and at first the result of the calculation is again something purely signitive, a certain combination of signs on paper." To solve a mathematical expression it is not necessary to fall back on the concepts, but it is sufficient to "combine signs with signs, replace sign-complexes by other sign-complexes, according to a rigidly rule-based procedure"[145].

A further criticism of Schröder, both in the review and in the *Logikvorlesung*, is that he did not understand why the calculus is applicable to different domains.

[143] *LV96*, 243–245.

[144] Repr. in: Husserl, *Aufsätze und Rezensionen*, 3–43.

[145] *LV96*, 247.

The calculus of classes does not provide a foundation for all the applications that it admits, but only codifies relations of subordination among sets. The analogies that allow for the application of the calculus are, according to Husserl, immediately evident. All the conceptual fields in which it makes sense to speak of set-theoretical inclusion, exclusion and coincidence can be interpreted by the same symbol system; for example the same algorithm is meaningful for classes of individuals, 'groups', equations, and in particular – what is most relevant in the present context – for concepts and judgements, according to their *Bedingtheitsverhältnisse*.[146]

2.8.2 On Propositional Inferences

Husserl's theory of propositional inferences is constituted by (i) certain basic laws (*Grundgesetze*) that are said to precede the constitution of any theory in general, (ii) a finite number of primitive axioms (*primitive Axiome*), and (iii) theorems (*Lehrsätze*) that are derived from (ii) by the rules of inference determined by (i).

Husserl provides the basic laws and axioms in the form of schemata: for every schema there are infinitely many possible instantiations that he calls "particularizations (*Besonderungen*)". As we will see later on, many of the "primitive axioms" that Husserl adopts recur in the logical calculi introduced more than twenty years later by Hilbert and Bernays.

As for the "derivations" of theorems, it does not seem too far fetched to claim that for Husserl the *proofs* in the theory of inferences are "formal objects" and that (as seems to be suggested in the passage quoted below) he would agree to identify them with what we would define as *formal proofs* in a logical calculus: finite sequences of formulae, each of which is either an axiom or is obtained from previous formulae by application of one of the primitive rules of inference or, more generally, by an inference "justified" by the theorems of implicational form demonstrated earlier. In this respect the following passage is very telling:

> An ideal theory of inferences would have to offer the following view. As mainstays (*Grundpfeiler*) we have certain primitive axioms, which cannot be derived from each other. Then follow the theorems, i.e. derived inferential laws. These derivations are again inferences or webs of inferences. But when we dissolve such a web into elementary inferences, then with the first theorem we will only get at those inferences that fall under the axiomatic principles as particularizations (*Besonderungen*). In the case of the second theorem, the inference that proves it can also have the form that has been shown to be valid by the first theorem, etc.[147] In short, whichever proof one may check and analyze in the

[146]Cp. Schröder Review and *LV96*, 242–248.

[147]In a formal proof (understood as a finite sequence of formulae that satisfies the previously mentioned requirements) the first formula has to be an axiom. By contrast, the second and all subsequent formulae are either axioms or derived from the previous ones by application of a rule of inference. But if in the proof we admit – as Husserl seems to do – the possibility of appealing to a previously proven theorem (without proving it again), the second formula can also be obtained through an instance of the first, the third through an instance of the first or the second, and so on.

> theory, one will always find in the series of axioms or previously demonstrated laws those that justify it.[148]

The resulting "hierarchical structure" of the theory of inferences therefore imposes as obvious requirements that none of the propositions that are to be proved is to be used in proofs (no vicious circles) and that "the derivations themselves do not fall under inferential laws that still have to be proved directly or indirectly through them".[149] However, Husserl allows for the case in which the inferential law to be proved is used in the *metatheoretical* inferential reasoning which we use to prove that law. It is really remarkable that Husserl has such a clear vision of the distinction between theory and metatheory as well as of the deductive machinery. This is very conspicuous when he writes:

> What would happen if we were to deduce an inferential law in such a way that the very law underlying the inference itself would be the one to be deduced? From law A we derive law B. But the principle of derivation granting the result B does itself presuppose B or has the form of B. In this case there is no vicious circle. We do have a circularity in the proof when we take ourselves to have proved a proposition as the consequence of another, which in its turn can be proved only by appeal to the proposition in question. Or in the case where we prove the truth of a proposition B from that of a proposition A, while A itself already contains B as explicit or hidden premise. In our case however the proposition to be proved is not a premise, but only provides the principle in accordance with which the proving inference proceeds.[150]

2.8.2.1 The General Principles

Let us now consider four "basic laws that must precede every [deductive] theory (*Grundgesetze, die allen Theorien vorhergehen müssen*)". The first of these basic laws is

(I) The principle that allows the inference "from the general to the particular *(vom Allgemeinen auf das Besondere)*".

For Husserl the fact that the axioms and rules of inference function as general forms or schemata whose instances are used in proofs must be made explicit as a law. Basic law (I) warrants that the general inferential laws with which deductive thought has to comply can be applied in the particular sciences. In proving one applies principles. For example if among the general principles we have commutativity (A + B = B + A), and we have proved a proposition containing $(a - b) + c$,

---

[148] *LV96*, 250.

[149] *LV96*, 249.

[150] *LV96*, 249–250.

then we can move to the proposition where $(a - b) + c$ is replaced by $c + (a - b)$, by application of that law.[151] "What does applying [a law] mean? It means inferring the truth of the particular case that falls under it from the truth of the general law.[152]

As second basic law we have

(II) The *modus ponens* of traditional logic, that is the rule A, A → B/B.

"If a proposition A is valid, and moreover it is valid that if A is true then also B is true, then also B is valid. Evidently this principle finds constant application."[153] (As ever so often in Husserl, 'being valid (*gelten*)' is just a stylistic variant of 'being true'.)

A further basic law that is "needed straightaway at the beginning of a theory" is given by

(III) The rules of conjunction introduction and conjunction elimination.

If proposition A is valid and so is proposition B, then A ∧ B is also valid, and *vice versa*. "If, for example, we want to connect various axioms or theorems we immediately need the proposition: if the general proposition A (*der allgemeine* [sc. Satz] *A*) is valid and the general proposition B is valid, then also the general proposition A and B (*der allgemeine* [sc. Satz] *A und B*) is valid, that is their conjunction is true, and *vice versa*".[154]

Husserl's fourth and final basic law is

(IV) The principle of distributivity of universal quantification over implication:

$$\forall \mathrm{x}(\mathrm{A}(x) \rightarrow \mathrm{B}(x)) \rightarrow (\forall x\mathrm{A}(x) \rightarrow \forall x\mathrm{B}(x)),$$

where x varies over an arbitrary class C.

Husserl formulates it as follows: "Let A and B be two general propositions, with a general relation to the objects u of some delimited class. Then we can say: assuming that it is valid that every u for which the proposition A is true, also makes the proposition B true, then it is certain that if A in general is true for each u, also B must be true for every u. For example if it is valid for every square that if it is divisible into two triangles then the sum of its internal angles must be equal to four right angles, then this is also valid: if every square is divisible in two triangles then every square has the sum of its internal angles equal to four right angles."[155]

---

[151] *LV96*, 250.

[152] *LV96*, 251.

[153] Loc. cit.

[154] Loc. cit.

[155] *LV96*, 251 f.

2.8.2.2 The Notation

Before discussing the formal structure that Husserl gives to the theory of propositional inferences (or laws), let us review his choice of primitive logical operators and their symbolic notation.[156]

As meta-variables for propositions Husserl uses the upper-case letters A, B, ...; as meta-variables for 'terms' the lower-case letters x, y, a, b, ...

- The conjunction of A and B is expressed by juxtaposition: AB
- The disjunction of A and B is expressed by the + sign: A + B
- The conditional with antecedent A and consequent B is expressed by €: A€B[157]
- The biconditional is expressed by the symbol =
- The negation of A is expressed by the sign 0 as an index: A_0
- Π and Σ represent the universal and the existential quantifier respectively

For a correct interpretation of Husserl's notation one should keep the following points in mind:

a. Husserl wants disjunction to be understood as *inclusive* ('*vel*').
b. The conditional, €, is used *both* for the *operation* that from two propositions A and B forms the new proposition 'if A then B' and for what Husserl takes to be the fundamental *relation* that can obtain between propositions: implication.[158] "A third elementary way to construct one proposition out of two is the hypothetical one: if A then B. But this mode of connection represents at the same time the fundamental form of the relation between propositions. The validity of A implies that of B, and thus both propositions are put in a relation from which originate peculiar relative properties for each: being a reason (*Grund*) and being a consequence (*Folge*)."[159]
c. The biconditional is defined *via* conditional and conjunction: A = B abbreviates (A€B)(B€A).
d. Besides the sign for negation (A_0 means the same as 'it is not true that A'), Husserl provides also a sign for the affirmation (the sign 1 as an index: A_1 is to mean the same as 'it is true that A'); but he observes that we do not need the latter symbol, since A and 'It is true that A' are always strongly equivalent.
e. Husserl also introduces (but never actually uses in the subsequent sketch of the "calculus") the symbol '!' that has essentially the same point as Frege's vertical, the *Urteilsstrich*: while A expresses a certain propositional content (in Frege's

---

[156]Most of the time we shall not use quotation-marks in our explanations when talking about symbols. Husserl's prose is very loose in this respect, too, but hopefully no confusions will be caused by this sloppiness.

[157]€ is the symbol Schröder uses to signify the relation of inclusion among classes.

[158]In this way logical laws with implicative form for Husserl also play the role of rules of inference. This explains the terminological wavering between "theory of propositional inferences" and "theory of propositional laws".

[159]*LV96*, 254.

Conceptual Notation A would be preceded by a horizontal stroke), A! expresses it with the force of a judgement. Frege represented this force by prefixing the judgement-stroke to the expression of a propositional content: ⊢ A). "A proposition can appear simply as proposition, or its corresponding truth can be meant at the same time. For example if we say that with the proposition 'God is just' is also given the proposition 'Evil is punished', if we say that when one is valid also the other is valid, then we do not affirm these propositions themselves as being valid. However, if we simply say 'God is just' we mean the truth 'God is just'. We will indicate the difference with an exclamation mark. Hence: A!"[160] The point of the symbol '!' has nothing to do with logical truth or validity, or with the fact that to prove something is to prove its truth. What Husserl wants to maintain (as Frege did before him) is that *the difference between merely thinking something without commitment as to its truth and thinking it with such a commitment must be formally expressible.*

f. About the quantifiers, specifically the universal quantifier Π (the existential quantifier Σ is introduced but never actually used in his "calculus"), Husserl remarks that the need to explicitly express universality (*Allgemeinheit*) by means of a specific sign is due to the fact that there are universal propositions in which not all of the terms (*Termini*) are variables (*Variablen*), that is 'bearers' of universality. "In every universal proposition the universality relates to certain variables, for example 'It is generally valid (*Allgemein gilt*) that a man is mortal'. Here the word 'a' is a sign for the variable. . . . In the arithmetical proposition 'an even and an odd number have as their sum an odd number' we have two variables. . . In these propositions, however, only these particular terms are bearers of universality (*Träger der Allgemeinheit*)".[161] Husserl then uses the notation $f(x\ y \ldots a\ b \ldots)$ to indicate a proposition in which the 'terms' $x, y, a, b, \ldots$, appear, so $\Pi_{xy\ldots} f(x\ y \ldots a\ b \ldots)$ means: for every $x, y, \ldots f \ldots$ holds (and analogously, $\Sigma_{xy\ldots} f(x\ y \ldots a\ b \ldots)$ means: there are $x, y, \ldots$ for which $f \ldots$ holds).

2.8.2.3 The Calculus

Finally we would like to reconstruct some of the sketches of a "calculus" of the "theory of propositional inferences" that Husserl presents. What we find in his lecture is not so much a systematic construction of a formal theory, but a series of very brilliant intuitions that point into this direction.

First of all, Husserl recapitulates the four basic laws that we already had occasion to discuss, but he partially modifies them, which again testifies to his continual rethinking of the issues. The modified versions are indicated by the letters α, β, γ and δ.

[160] Loc. cit.

[161] *LV96*, 253.

(α) The first principle, "the inference from the general to the particular", is reformulated as a principle concerning the relations between genus and species. Indirectly, the distinction between the notion of *belonging* (of an element to a class) and that of *inclusion* (among classes) is taken into account. Husserl points out that we must now consider not the inference from the general to the particular *tout court*, but the "inference from the general to the particular *that is subordinated to it*", and that this particular, far from being a singular case (*ein Einzelnes*), must also be something general. "When we apply a proposition about all quadrangles to squares in particular, then the particular here is of a different kind than when we apply a proposition that is valid for all men to Socrates. In the first case the particular is itself something general, in the second case it is not. For our purposes we will consider the principle only in that sense where the particular is itself general".[162] Therefore the "official" formulation of principle (α) is the following: "if a proposition *f* is generally valid for any *u, v, ... z*, and at the same time it is valid that, in this series, every *u'* is a *u*, every *v'* is a *v*, and finally every *z'* is a *z*, then the proposition *f* is also valid in particular for every *u', v', ... z'*".[163]

It is important to notice that among the applications of this principle that Husserl mentions, we find not only (as was to be expected) inferences of the kind "if a proposition is valid in general for every quadrangle, then it is also valid for every square, for every rectangle, for every trapeze etc.", but also inferences that instantiate to certain classes of expressions (e.g. to all the expressions of a certain logical form) a law that is valid in general for every expression. "This principle allows us to apply every law that is valid for propositions in general to arbitrary combinations, disjunctions or hypothetical connections of propositions, and this in a completely general way, so that the resulting propositions have again the character of laws, of formulae".[164] The effective application of principle (α) in the remaining pages to be taken into consideration will be precisely of this kind: it is used as if it were a kind of *substitution rule for the logic of connectives*. Actually, the "calculus" that we are discussing is a calculus for the logic of connectives, not for that of the quantifiers. (The latter, or rather its monadic fragment, will be the topic of the theory of conceptual inferences.)

(β) As second principle we find again the rules of conjunction introduction and of conjunction elimination, extended to "universal closures": "if *f* is generally valid [*gilt allgemein*, i.e. holds of everything] and *g* is generally valid, then also the united proposition *f and g* is generally valid, and indeed with respect to all the variables that occur in *f* as well as in *g*, and *vice versa*".[165] Husserl will then

[162] *LV96*, 254.

[163] Loc. cit.

[164] *LV96*, 254 f.

[165] *LV96*, 255.

use it at times as law rather that as rule; in his notation (where ΠA is the "universal closure" of A): Π(AB) = ΠAΠB.

(γ) The third and fourth principle of those introduced above are "fused" in the rule of *modus ponens* extended to "universal closures": "if the hypothetic proposition 'if *f* then *g*' is generally valid, we must conclude that, if *f* is generally valid, also *g* is generally valid".[166] In Husserl's notation: from Π(A€B) and ΠA follows ΠB.

Notice that Modus ponens can be found among the general principles in the guise of a rule, while we also find it as implicational law among the axioms of the theory of inferences.[167]

(δ) Finally we find a principle that Husserl calls "trivial" and that is a variant of (β), this time formulated as a law having the form of a biconditional and with the explicit assertion of the truth of the premises and of the conclusion: "if A is true and B is true then the proposition A and B is also true, and *vice versa*. $A_1B_1 = (AB)_1$".[168]

The *logical* axioms are the following,[169] grouped by Husserl according to the "connectives" they contain. In the first group we notice the law of *modus ponens* (I), the law of *transitivity* (II), the *lattice-theoretical* laws for conjunction, including commutativity (III, IV and V). The second group consists of the law of *modus tollens* or the law of contraposition in imported form (VIII), the laws of *non-contradiction* and of *double negation* (IX, X), and one of the two *De Morgan* laws (XI).

Group 1 – conditional and conjunction:

	[Husserl's notation]	["translation"]
I.	A(A€B) € B,	$A\wedge(A\rightarrow B)\rightarrow B$;
II.	(A€B)(B€C) € (A€C),	$(A\rightarrow B)\wedge(B\rightarrow C)\rightarrow(A\rightarrow C)$;
III.	(M€A)(M€B) € (M€AB),	$(M\rightarrow A)\wedge(M\rightarrow B)\rightarrow(M\rightarrow A\wedge B)$;
IV.	AB € A,	$A\wedge B\rightarrow A$;
V.	AB € BA,	$A\wedge B\rightarrow B\wedge A$;
VI.	(AB€C)A € (B€C),	$(A\wedge B\rightarrow C)\wedge A\rightarrow(B\rightarrow C)$;
VII.	(A€B) € (A€AB),	$(A\rightarrow B)\rightarrow(A\rightarrow A\wedge B)$.

Group II – disjunction and negation (plus conditional and conjunction):

VIII.	$(A€B)B_0$ € A_0,	$(A\rightarrow B)\wedge\neg B\rightarrow\neg A$;
IX.	$(AA_0)_0$,	$\neg(A\wedge\neg A)$;
X.	$(A_0)_0 = A$,	$\neg\neg A\leftrightarrow A$;
XI.	$(A+B)_0 = A_0B_0$,	$\neg(A\vee B)\leftrightarrow\neg A\wedge\neg B$;

[166]Loc. cit.

[167]As is the case for example in the *Principia Mathematica* of Russell and Whitehead.

[168]Loc. cit.

[169]We reproduce them here in Husserlian notation as well as in a transcription using current symbolism. With ∀[A] we indicate the universal closure of A.

To these axioms Husserl then adds two concerning quantification (they are never used in the proofs). But his use of small Greek letters to designate them suggests that Husserl takes them to be on the same level as the fundamental principles, to be added to the four mentioned in the beginning):

ε)	$\Pi A + \Pi B \in \Pi(A + B)$[170]	$\forall[A] \lor \forall[B] \to \forall[A \lor B]$;
ι)	$\Pi(A_0) \in (\Pi A)_0$[171]	$\forall[\neg A] \to \neg\forall[A]$.

On the base of these axioms and using the principles (α) – (δ), Husserl proves (or affirms the easy provability of) a whole list of *theorems* (*Lehrsätze*) that we provide[172] in its entirety.

Lemmata that follow from the axioms of group I:		
1.	$(A\leftrightarrow B) \to (A\to B)$,	
2.	$(A\leftrightarrow B) \to (B\to A)$,	
3.	$(A\leftrightarrow B) \leftrightarrow (B\leftrightarrow A)$,	
4.	$A\land B \to B$,	
5.	$A\land B \leftrightarrow B\land A$,	
6.	$(\forall[A]\land\forall[B])\land\forall[C] \to \forall[(A\land B)\land C]$,	
7.	$\forall[A]\land\forall[A\to B] \to \forall[B]$,	
8.	$\forall[A]\land \forall[B]\land\forall[A\land B\to C] \to \forall[C]$,	
9.	$A\land(B\land C) \to (A\land B)\land C$	(*associativity of* ∧),
10.	$(A\to B)\land(A\leftrightarrow C) \to (C\to B)$,	
11.	$(A\to B)\land(B\leftrightarrow C) \to (A\to C)$,	
12.	$\forall[A\land B\to C]\land\forall[A] \to \forall[B\to C]$,	
13.	$A \to (B\to A)$	(*a fortiori*),
14.	$(A\to B) \to (A\land C \to B\land C)$,	
15.	$(A\to B)\land(C\to D)\to (A\land C \to B\land D)$,	
16.	$(A\to B)\land(C\leftrightarrow D)\to (A\land C \to B\land D)$,	
17.	$(A\leftrightarrow B)\land(C\leftrightarrow D)\to (A\land C \leftrightarrow B\land D)$,	
18.	$(A\land B\to C) \to (A\to(B\to C))$	(*exporting the premise*),
19.	$(A\to(B\to C)) \to (A\land B\to C)$	(*importing the premise*),
Lemmata that follow from the axioms of groups I and II:		
20.	$(A\to B) \to (\neg B\to\neg A)$	(*contraposition*),
21.	$(A\leftrightarrow B) \to (\neg B\leftrightarrow\neg A)$	
22.	$A\lor\neg A$	(*excluded middle*),
23.	$(A\lor B) \leftrightarrow (B\lor A)$	(*commutativity of* ∨),

(*continued*)

[170]In the text we find $\Pi(A + B) \in \Pi A + \Pi B$, but this is clearly a mistake or a trasnscription error; besides, Husserl explicitly points out: "only one half is valid" (*LV96*, 259).

[171]This is what we find in the text, and surely it is a logical law. However, it could be, as above, a mistake or transcription error (per the logical law $\Pi(A_0) \in (\Sigma A)_0$).

[172]We do not use Husserl's notation here; moreover we correct where necessary some errors that are present in the text regarding the progressive numeration (errors that indicate various stages of rewriting and rethinking these pages).

Lemmata that follow from the axioms of groups I and II:		
24.	$A\vee(B\vee C) \rightarrow (A\vee B) \vee C$	(*associativity of* $\vee$),
25.	$(A\rightarrow B) \rightarrow (A\vee C \rightarrow B\vee C)$[173],	
26.	$(A\rightarrow B)\wedge(C\rightarrow D)\wedge(A\vee C) \rightarrow B\vee D$,	
27.	$A \rightarrow A\vee B$	(*disjunctive weakening*),
28.	$(A\rightarrow C)\wedge(B\rightarrow C) \rightarrow (A\vee B\rightarrow C)$,	
29.	$(A\vee B)\wedge C \rightarrow (A\wedge C) \vee (B\wedge C)$	(*distributivity*),
30.	$(A\rightarrow B) \leftrightarrow\neg(A\wedge\neg B)$	(*Chrysippus' law*).

As we anticipated, Husserl provides a (more or less detailed) proof for some of these lemmata. Let us consider, as an example, four of the most representative ones, presenting them in the current style (the "justifications" in right column are those that can be desumed from the text) and, in two cases, we will also report the proof in Husserl's words.

Notice, in particular, the peculiar usage (neither always explicit nor always consequential) of the principles (β) – (δ).

Proof of (5): $A\wedge B = B\wedge A$.

1.	$\forall[A\wedge B \rightarrow B\wedge A]\wedge\forall[B\wedge A \rightarrow A\wedge B] \rightarrow \forall[(A\wedge B \rightarrow B\wedge A)\wedge(B\wedge A \rightarrow A\wedge B)]$	(β)
2.	$\forall[A\wedge B \rightarrow B\wedge A]$	V
3.	$\forall[B\wedge A \rightarrow A\wedge B]$	V
4.	$\forall[A\wedge B \rightarrow B\wedge A]\wedge\forall[B\wedge A \rightarrow A\wedge B]$	(δ): 2,3
5.	$\forall[(A\wedge B \rightarrow B\wedge A)\wedge(B\wedge A \rightarrow A\wedge B)]$	(γ): 1,4
6.	$A\wedge B = B\wedge A$	5, def.

Proof of (6): $(\forall[A]\wedge\forall[B])\wedge\forall[C] \rightarrow \forall[(A\wedge B)\wedge C]$.

1.	$(\forall[A]\wedge\forall[B]) \wedge\forall[C] \rightarrow \forall[A]\wedge\forall[B]$	IV
2.	$(\forall[A]\wedge\forall[B]) \rightarrow \forall[A\wedge B]$	(β)
3.	$(\forall[A]\wedge\forall[B]) \wedge\forall[C] \rightarrow \forall[A\wedge B]$	II: 1,2
4.	$(\forall[A]\wedge\forall[B]) \wedge\forall[C] \rightarrow \forall[C]$	L 4
5.	$(\forall[A]\wedge\forall[B])\wedge\forall[C] \rightarrow \forall[A\wedge B]\wedge\forall[C]$	III: 3,4
6.	$\forall[A\wedge B]\wedge\forall[C] \rightarrow \forall[(A\wedge B)\wedge C]$	(β)
7.	$(\forall[A]\wedge\forall[B])\wedge\forall[C] \rightarrow \forall[(A\wedge B)\wedge C]$	II: 5,6

Proof of (18): $(A\wedge B\rightarrow C) \rightarrow (A\rightarrow(B\rightarrow C))$.

«Principle VI itself has the form AB € C. Applying VI to it, we obtain:

$$\underbrace{((AB€C)A€(B€C))}_{A}\underbrace{(AB€C)}_{B} € \underbrace{(A€(B€C))}_{C}$$

But this proposition has again the form AB € C. At the same time it also is a law. The "A" in it is again a law. If we then apply 12 $[\forall[A\wedge B\rightarrow C]\wedge\forall[A] \rightarrow \forall[B\rightarrow C]]$,

[173]The text erroneously has a biconditional instead of the main conditional.

we immediately obtain the proposition. On the left side we indeed have two correct propositions. Therefore the right side also is a correct proposition (*modus ponens*)».[174]

We can render this as follows:

1.	$((A\wedge B\rightarrow C)\wedge A \rightarrow (B\rightarrow C))\wedge((A\wedge B\rightarrow C)) \rightarrow (A \rightarrow (B\rightarrow C))$	VI
2.	$(A\wedge B\rightarrow C)\wedge A \rightarrow (B\rightarrow C)$	VI
3.	$1\wedge 2 \rightarrow ((A\wedge B\rightarrow C) \rightarrow (A \rightarrow (B\rightarrow C)))$	VI
4.	$1\wedge 2$	(β):1,2
5.	$((A\wedge B\rightarrow C) \rightarrow (A \rightarrow (B\rightarrow C))$	(γ): 3,4

Proof of (30): $(A\rightarrow B) \leftrightarrow\neg(A\wedge\neg B)$.

"One half we can prove thus:

By 14: (A€B) €, multiplying both sides, (AB_0 € BB_0). Let us consider the right side: by VIII: (AB_0 € BB_0)(BB_0)$_0$ € (AB_0)$_0$; hence by 12:

$$(AB_0€BB_0)€(AB_0)_0.$$

Combining this, on the base of II, with the previously derived proposition, we get:

$$(A€B)€(AB_0)_0.$$

Proof in the other direction (much more complex):

$((A_0)_0\ B_0)_0$ € $A_0 + B$. Substituting in XI^β A_0 for A. But by X: $(A_0)_0 = A$. Therefore $(A_0)_0B_0 = AB_0$. Hence (3°) also the negations of both sides equal. And clearly we can substitute equivalence. Therefore $(AB_0)_0$ € $A_0 + B$; $A(A_0 + B)$ € $AA_0 + AB$. But $M + aa_0 = M$; hence € AB€B; $A(A_0 + B)$ € B; $A_0 + B$ € (a € b)".[175]

In our notation this can reproduced as follows:

From left to right:

1.	$(A \rightarrow B) \rightarrow ((A\wedge\neg B) \rightarrow (B\wedge\neg B))$	L 14
2.	$((A\wedge\neg B) \rightarrow (B\wedge\neg B))\wedge\neg(B\wedge\neg B) \rightarrow \neg(A\wedge\neg B)$	VIII
3.	$\neg(B\wedge\neg B)$	IX
4.	$((A\wedge\neg B) \rightarrow (B\wedge\neg B)) \rightarrow \neg(A\wedge\neg B)$	L 12 (and L 6): 2,3
5.	$(A \rightarrow B) \rightarrow \neg(A\wedge\neg B)$	II: 1,4

[174] *LV96*, 258.

[175] *LV96*, 259–260 (the *lower-case* letters are in the text).

From right to left:

1.	$\neg(\neg\neg A\wedge\neg B) \rightarrow \neg A\vee B$	XI[176]
2.	$\neg\neg A \leftrightarrow A$	X
3.	$\neg\neg A\wedge\neg B \leftrightarrow A\wedge\neg B$	L 14: 2
4.	$\neg(\neg\neg A\wedge\neg B) \leftrightarrow \neg(A\wedge\neg B)$	L 21: 3
5.	$\neg(A\wedge\neg B) \rightarrow \neg A\vee B$	II (and L 2): 1,4
6.	$A\wedge(\neg A\vee B) \rightarrow (A\wedge\neg A)\vee(A\wedge B)$	L 29
7.	$(A\wedge\neg A)\vee(A\wedge B) \leftrightarrow (A\wedge B)$	"$M + aa_0 = M$"[177]
8.	$A\wedge(\neg A\vee B) \rightarrow (A\wedge B)$	II (and L 1): 6,7
9.	$A\wedge B \rightarrow B$	L 4
10.	$A\wedge(\neg A\vee B) \rightarrow B$	II: 8,9
11.	$(\neg A\vee B) \rightarrow (A\rightarrow B)$	L 18: 10
12.	$\neg(A\wedge\neg B) \rightarrow (A\rightarrow B)$	II: 5, 11

One may very well wonder why the list of theorems contains neither the law *of identity*, $A\rightarrow A$, nor the two laws of idempotence for conjunction and disjunction ($A\wedge A \leftrightarrow A$; $A\vee A \leftrightarrow A$). Husserl explains the reason for this absence towards the end of the section that we are analyzing. Strictly speaking, he maintains, a conditional in which antecedent and consequent coincide is *meaningless*, in so far as there cannot obtain "an objective relation of *conditionality* (*objektives Verhältnis der Bedingtheit*)" between a proposition A and A itself. If we utter a natural language sentence that apparently has precisely this form (e.g. "If I have ordered something, then I have ordered something"[178]), we really mean (*meinen*) something quite different from a conditional relation (in the example, the speaker means that he does not retract the orders that he has given, etc.).[179] In complete analogy, also disjunctions and conjunctions with identical members ($A\vee A$, $A\wedge A$) are, strictly speaking, meaningless: "When we say: '2×2 is 4 and 2×2 is 4', then we have repeated the very same proposition, but we have not performed a conjunction of objective propositions".[180]

How can this be reconciled with the need to assure that every formal expression remains meaningful under every substitution? Husserl's way out consists in *assigning* a meaning to these formulae, too, and precisely a *conventional* meaning. Replacing in lemmata n. 30 (*Chrysippus'* law) B by A, we obtain: $(A\rightarrow A) \leftrightarrow \neg(A\wedge\neg A)$, and hence we can posit *by convention* $A\rightarrow A =_{df} \neg(A\wedge\neg A)$. Analogously, we can posit by convention $A\wedge A =_{df} A$, $A\vee A =_{df} A$.[181]

It is interesting to observe that Husserl sees in this solution an analogy with arithmetic: "we can operate computationally (*rechnerisch*) without worrying

[176] It is unclear what is meant by 'β' in 'XI^{β}'.

[177] $A\vee(B\wedge\neg B) \leftrightarrow A$ is not listed among the theorems.

[178] *LV96*, 260.

[179] Cp. Bolzano on Pilate's "What I have written I have written": *WL* II, §148, 85, end of note 1.

[180] *LV96*, 260.

[181] In his last published paper, "Compound Thoughts" (1923), Frege maintains that $A\vee A$ and $A\wedge A$ express the same thought as the plain A and that $A\rightarrow A$ expresses the same thought $\neg(A\wedge\neg A)$: see p 49 and 50 (original pagination).

whether or not the members of the connections are identical or not. These conventions here have the same function as certain analogous ones in arithmetic, thanks to which 1 and 0, which in the original sense are not numbers, are adjoined to the numbers by certain conventions... Then we can also understand the arithmetical formulae in such a general way that the letters could just as well stand for 0 and 1 as for proper numbers".[182]

Husserl's concluding remarks also deserve attention. To the question whether it would be possible "to express simple truths in conditional form", Husserl gives a positive answer, and he is right. It suffices to observe, he says, that all contradictions ($A\wedge\neg A$, $B\wedge\neg B$, ...) are logically equivalent and that, if we indicate any of these with '0', we can prove (in the "calculus" we described) $\neg A \leftrightarrow (A\rightarrow 0)$, for every A. Analogously, given that we have $A\vee\neg A \leftrightarrow B\vee\neg B \leftrightarrow C\vee\neg C \leftrightarrow \ldots$, by indicating any of these expressions with '1' we also obtain $A \leftrightarrow (1\rightarrow A)$, for every A.

2.8.3 On Predication and Conceptual Inferences

The topics of Husserl's theory of conceptual inferences are the "laws that are founded in the concepts of *object* and *concept*, that are valid for *objects* and *concepts* as such".[183] More precisely, the theory aims at characterizing the laws that are based firstly on the fundamental relation, linguistically expressed by predication, that obtains between an object and a concept just in case that object stands under that concept (Frege called this relation *subsumption*), and secondly on the relation of *subordination* among concepts.

Let us first have a brief look at Husserl's theory of predication. It is to be found in the second part of his *Logikvorlesung* of 1896 entitled "On Propositions (*Von den Sätzen*)". Husserl distinguishes here two canonical forms of sentences to which, he maintains, all sentences of natural languages are, in principle, reducible: the categorical form (*kategorische Form*),[184] subject – copula – predicate (in symbols: 'S is P') and the existential form (in symbols: 'S is'). In this respect Bolzano's reductionism was even more radical: he assumed that all propositions could be expressed by sentences of the categorical form, or rather of the form 'S has the property of being P'.[185] In our context only the categorical form is of interest. Unlike traditional logicians Husserl does not take it to express general propositions – where both the subject *S* and the predicate *P* are general terms ('a man', 'white', 'rational',...). Rather, the subject *S* is always intended to be a singular term ('the Sun', 'Socrates',

[182] *LV96*, 261.
[183] *LV96*, 262 (my italics).
[184] *LV96*, 163–165.
[185] *WL* II, §127, 9.

...), that is, a term that purports to denote an individual. So sentences of the form 'S is P' have the following structure:

singular term – copula – general term.

From the ontological point of view, this corresponds to the relation of belonging (*zukommen*) that holds between a property and an object (in this order) if the sentence expresses a truth.

One may very well wonder where the copula is to be found in a sentence like "Socrates walks". Aristotle tried to uncover a copula even in such sentences: "It makes no difference whether we say of a man that he walks (βαδίζει), or whether we say that he is walking (βαδίζων ἔστιν)."[186] (When one renders Aristotle's *constructio periphrastica* in this way, one better forgets about the English progressive aspect.) Bolzano took the same line:

> Each inflected verb that is different form the word 'is' can be replaced, without any essential change of meaning, by the combination of 'is' with the (present tense) participle derived from that verb. '*A thut*' is tantamount to '*A ist thuend*'.[187]

In this respect Frege strategy seems to be more reasonable because it spares us a clumsy paraphrase:

> Often the word "is" serves as copula ... As such it can sometimes be replaced by a verb-ending. Compare, for example, "this leaf is green" and "this leaf greeneth".[188]

Husserl follows suit:

> When I say "the flower withers", the word "withers" contains both the expression for the pertinent P and something which corresponds to the word "is".[189]

As to the specific function of the copula, Husserl regards it as an unsaturated component, "( ) is ( )", which becomes a saturated whole, a sentence, once the places for the subject and for the predicate are suitably filled. As in Frege, the relation between object and concept is not invertible: the subject can never become a predicate; the predicate cannot become a subject without a previous essential transformation (nominalization). The relation between object and concept is primitive and cannot be defined. Husserl takes the schematic form "S is P" to have a content, which is also indefinable: it expresses the relation of standing under (*stehen unter*) that holds between an object and a concept. (The term is Bolzano's. Frege has it that an object "falls under" a concept.) As a matter of fact Husserl's claim here is similar to (though also, as we shall see, importantly different from) the thesis which Frege puts forward in his paper "On Concept and Object" (1892):

[186]Aristotele, *De Int.* 12: 21b9–10.

[187]*WL* II, §127, 10.

[188]Frege 1892, 194.

[189]*LV96*, 144.

> What I call here the predicative nature of concepts is just a special case of the need of supplementation, the unsaturatedness, that I gave as the essential feature of a function in my work 'Function and Concept' [1891].[190]

The difference between object and concept is an absolute one.

> A concept ... is predicative. For it is the *Bedeutung* of a grammatical predicate. By contrast, a name of an object, a proper name, is quite incapable of being used as a grammatical predicate.[191]

But we should not overlook a fundamental divergence between Husserl's (as of 1896) and Frege's conception. According to Frege, the unsaturated component in a sentence of the form 'S is P' is what *he* calls a predicate, namely an instance of '( ) is P'. By contrast, Husserl follows the tradition in classifying the general term 'P' as predicate, and for him the unsaturated component is the copula. By taking the copula to be more than just a component of a Fregean predicate Husserl can account for an interesting feature of non-nominal quantification in natural languages. When we move from 'Socrates is wise' to 'Socrates is something', in symbols: '$\exists\phi$ (Socrates is ϕ)', we quantify into the position of the general term 'wise', that is to say, into the position of the *Husserlian* predicate.[192] It is highly doubtful whether there is something like quantification into the position of a Fregean predicate in a natural language like English. (Of course this is no objection against Frege's understanding of second-order quantification in his "logically perfect language [*Begriffsschrift*]".)

Let us now turn to Husserl's theory of conceptual inferences. While its goal is expressed quite clearly, the sketch of a formalization that Husserl presents is far less detailed than the one of the theory of propositional inferences. Actually there are just a few hasty and concise remarks that take only a bit more than one page. From the strictly formal point of view we can say, using a terminology that is not Husserl's, that the result is essentially nothing but a part of the "theory of classes".

On the notational level, Husserl uses the symbols:

- Γ, Δ, ... as variables for objects (individuals)
- a, b, ... A, B, ..., α, β ... as placeholders for monadic general terms
- ε for the binary relation of predication
- so 'ΓεA' is the formal representation of a subsumption "the object Γ is A", that is "Γ has the property of being A"[193]

For the relation of subordination between concepts, the other fundamental relation that is a topic of the theory, Husserl's notation is the sign € (that in the theory of inferences is the connective of the material conditional) with a dot on top

190 Frege 1892, 197 n.

191 Frege 1892, 193 and n.

192 Cp. Künne 2003, 65 f.

193 Husserl emphasizes that, even if Γ is considered as an indeterminate object (an *Etwas*), 'Γε a' is not to be confused with the existential statement "Something is a", for which he suggests the formal notation 'Σa'.

of it. For typographical reasons we shall use the standard symbol for inclusion, '$\subseteq$'. So '$A \subseteq B$' expresses the quantified conditional *if something is A, then it is B* or, equivalently, the universal affirmation *all A are B*. It is not necessary – Husserl observes – to take the relation of subordination as primitive, since it can be defined with the help of *individual* quantification, the predication relation and the material conditional:

$$A \subseteq B =_{df} \forall \Gamma (\Gamma \varepsilon A \rightarrow \Gamma \varepsilon B).$$[194]

Without explicit acknowledgement Husserl uses the symbol '$\subseteq$' not only for the relation of inclusion but also for an operation on concepts that associates with the concepts A and B the *concept* $A \subseteq B$ under which all and only those objects Γ fall that are B under the hypothesis that they are A. So the symbol '$\subseteq$' is unfortunately ambiguous.

Finally, Husserl uses the notation 'AB' (without having introduced it explicitly with this new reading) to indicate the concept of something that is A as well as B, that is extensionally, the intersection of the classes A and B.[195] (In the theory of propositional inferences such a juxtaposition of letters was used to indicate the conjunction of two propositions.)

As regards the axiomatization of the theory, it is not possible – for the reasons given above – to reconstruct precisely what Husserl had in mind. Apparently he was thinking about an axiomatization based on that of the theory of propositional inferences, expanded with the appropriate *Grundformeln* and with a "conceptual reading" of the fundamental principles. Three of the only four "conceptual laws" he mentions explicitly in our text:

$$(B \subseteq C) \wedge (A \subseteq B) \rightarrow (A \subseteq C), \tag{1}$$

$$(A \subseteq B) \wedge (A \subseteq C) \rightarrow (A \subseteq BC), \tag{2}$$

$$(A \subseteq (B \subseteq C) \rightarrow ((A \subseteq B) \rightarrow (A \subseteq C)), \tag{3}$$

are in fact, as Husserl points out, the "conceptual" counterpart of the propositional principles (α), (β) and (γ).

(1) is a counterpart to the syllogistic *modus barbara* and corresponds to the "principle of substitution" (α); (2) corresponds to the "principle of adjunction" (β), that is $\Pi A \Pi B = \Pi(AB)$; (3) corresponds to the "generalized *modus ponens*" (γ), that is $\Pi(A \text{€} B)$ € $(\Pi A$ € $\Pi B)$. Clearly, in (3) the second occurrence of the symbol $\subseteq$ denotes the operation on concepts mentioned above and not the relation of inclusion.

---

[194]*LV96*, 263.

[195]Husserl's use of his notation is somewhat unstable: instead of 'AB' he sometimes writes 'A_B'.

The fourth law is:

$$(A \subseteq (B \subseteq C) \leftrightarrow (AB \subseteq C), \tag{4}$$

the conceptual counterpart of the law of "importing/exporting the antecedent". Husserl also provides a sketch of a proof of (3) starting from (4), (1), (2), using propositional inferences. Indeed, by substitution, from (1) we obtain:

$$(AB \subseteq C) \wedge (A \subseteq AB) \rightarrow (A \subseteq C);$$

then, using (2),

$$(AB \subseteq C) \wedge (A \subseteq B) \rightarrow (A \subseteq C),$$

and by propositional inference (exporting):

$$(AB \subseteq C) \rightarrow ((A \subseteq B) \rightarrow (A \subseteq C)),$$

From this and (4), by transitivity of the implication, follows (3).

Husserl recognized the close "parallelism" between *conceptual* and *propositional* inferences and consequently between the respective axiomatic calculi. So he must have had a clear awareness of what Boole already knew and what only Schröder had precisely spelt out: the *double interpretation* of the "logical calculus", that is of the "Boolean calculus", as a logic of classes and as a logic of propositional connections:

> A first part of the logical calculus is constituted ... by *calculation with concepts* (*Rechnung mit den Begriffen*); through this it is possible to execute those inferences whose premises and conclusions are '*judgements of the first class*', i.e. judgements in which something is stated about things themselves – normally, categorical judgements.
>
> The second part comprises *calculating with judgements* (*Rechnen mit den Urteilen*) ... In this part judgements are made about our affirmations about things. These judgements concern the way in which the truth or untruth of certain affirmations turns out to be dependent on that of others, hence they concern relations that standardly find their linguistic expression in conditional propositions, in disjunctive or hypothetical judgements, which we want to call, following Boole, '*judgements of the second class*'.
>
> While in both parts calculation proceeds in accordance with the same laws, in each of them it is only the interpretation of the formulae that changes.[196]

In his reply to Schröder's review of his *Begriffschrift* Frege noted two weaknesses of the Boole-Schröder type of logic. The first one consists in its inability to represent deductive relations between judgements of the first and judgements of the second class. Frege writes:

---

[196]Schröder 1877, 1.

> The real difference [between Boole's logic and my *Begriffsschriftt* which Schröder does not notice at all] is that I avoid such a division into two parts ... and give a homogeneous presentation of the whole. In Boole the two parts run alongside one another, so that one is like the mirror image of the other, but for that very reason stands in no organic relation to it.[197] Each transition from a judgement of one kind to one of the other is cut off, though such transitions very often occur in real life.[198]

Boole himself gives (i) "All inhabitants are either Europeans or Asiatics" as an example of a judgement of the first class and (ii) "Either all inhabitants are Europeans, or all inhabitants are Asiatics" as an example of a judgement of the second class.[199] Obviously, the argument '(ii) therefore (i)' is deductively correct, but only in Frege's logic this transition can be formulated and legitimized.

Furthermore, judgements of the first class correspond to monadic quantification in Frege's system. In Boole's algebra of classes *multiple* quantifications cannot be adequately represented. One of the glories of the integrated Fregean system is that it is able to make sense of multiple quantifications which pervade arithmetic ('For *each* prime number *there is at least one* that is greater than it').[200]

2.9 Concluding Remarks

In this chapter we have focused on Bolzano's sadly neglected role in Husserl's work from, say, 1896–1900. We tried to bring to light where Husserl is working with Bolzanian notions, though not always explicitly acknowledging it, and where his reflections – as in the case of the so called *threefold stratification of logic* – go one step further. Many Bolzanian issues are discussed in the logic lecture of 1896 and in the *Prolegomena*: (i) Bolzano's conception of logic as a *Wissenschaftslehre* or "science of all possible sciences", (ii) Bolzano's rediscovery of the conception of an objective relation of dependency among the homogeneous truths which make up a specific science, (iii) Bolzano's theory of variation as providing a basis for Husserl's theory of semantic categories, (iv) Bolzano's notion of derivability (*Ableitbarkeit*) by means of which it is possible to cope both with logical inference and with enthymematic inference, (v) Bolzano's concept of a semantic relation of consecutivity (*Abfolge*) between pure conceptual truths.

Of particular interest to us was a contrast between two different approaches to the concept of proof resp. of theory. In classical logic this contrast is exemplified by the two contrasting (but equivalent) paradigms of logical calculi: Frege-Russell-Hilbert (the axiomatic approach: laws are privileged) vs. Gentzen (no logical laws, only logical rules). The former is an approach of a *synthetic* kind, the latter of an

197 Frege 1881, 15; transl. 14.

198 Frege 1882, 100.

199 Boole 1847, 58–59.

200 For the last two points cp. Künne 2009, Chapter 5 and the literature registered there.

analytical kind. We have considered the concept of an *etiological proof* resp. of a *privileged form of theory*, in which every derivation answers the question *why* something is true. Our intention was to show that Husserl, following Bolzano's footsteps, is fully aware of the problem to which an analytical approach tries to give an answer, that is the 'problem of *Methodenreinheit*': in order to arrive at a certain proposition we should not use concepts that are extraneous to it.

However, in his *Logikvorlesung '96* Husserl builds up a calculus in the synthetic-axiomatic style. In Appendix 5 we tried to give a formal reconstruction of it and emphasized a specific difficulty. Husserl explicitly introduces the symbol "$\rightarrow$" (in Schroeder's notation (€)) as a sentence-forming operator, that is as a connective. But at the same time he stresses that "$\rightarrow$" also represents the relation of "inferability" between sentences. This means that "$\rightarrow$" has also an operational algorithmic reading: by knowing that $A\rightarrow B$ is a logical law we also know that whenever A is true then B must be true. Hence conditional laws can also be used as valid inference rules, that is derivation rules. (This is what Husserl in fact does, although only tacitly, in his development of the theory of propositional inferences.) Here, however, we stumbled over a difficulty: Husserl characterizes the inferential reading of "A$\rightarrow$ B" as the relation by which A is the ground of B and B is a consequence of A. But then the interreducibility (already known to the Stoics) of the notions of conditional laws and valid inference schemata obviously breaks down: for example the conditional law $A \wedge B \rightarrow A$ says, in the relational reading, that we can infer A from $A \wedge B$. This makes sense if we understand '$\rightarrow$' as a sign for *derivability*, but it makes no sense in the stronger reading which would require $A \wedge B$ to be a ground for A; since a ground has to be conceptually simpler than its consequence. In a more general form our perplexity can be formulated as follows: why does Husserl explicitly adopt in his conception of *Begründung* the Bolzanian idea of an inference rule of the kind *Abfolge* and then goes on to build up an axiomatic-synthetic style calculus and work with valid inferences of the kind *Ableitbarkeit*? Our answer was that, though accepting the requirements of the *Methodenreinheit*, Husserl is simultaneously trying to give a model for derivability that is as close as possible to what usually happens in current mathematical practice. A question arises at this point: isn't this what normally happens in various normalizations results given for example for calculi of natural deduction or in sequent calculi? For example in the case of sequent calculi, by allowing the cut-rule we allow in our derivations also formulae that are *not* subformulae of the conclusion; but we know, thanks to the normalization (cut elimination) theorem, that our derivation can be always transformed into a derivation of the same conclusion in which there are no applications of the cut rule, and so in a derivation satisfying the so-called subformula principle.

Chapter 3
The Imaginary in Mathematics

3.1 Introduction

The main aim of this chapter is to present both an "analytical" and an "internal" reading of Husserl's *Doppelvortrag*, the lecture he presented to the *Mathematische Gesellschaft* in Göttingen in the winter semester of 1901/02.[1] We will propose a formal reconstruction of some of the main results contained in the *Doppelvortrag* wherever a mathematical definition of the concepts helps to clarify Husserl's own ideas.

Before considering the lecture's content, clarification of the status of the text is in order. All that remains of the *Doppelvortrag* is contained in the manuscript with signature K I 26 conserved in the Husserl-Archives Leuven, which bears the title (written by Husserl in pencil) "ad Vortrag *Göttingen 1901*." Of the surviving fragments only a part can be legitimately considered as belonging to the final version of the *Doppelvortrag*, while the rest must be attributed to its preparatory stages. We know that Hilbert urged upon Husserl to publish the *Doppelvortrag* and that Husserl was occupied with the elaboration of the definitive text during the Christmas holidays 1901/1902.[2] Husserl did not publish the *Doppelvortrag* then, nor did he ever reconsider the possibility of a publication, though on several occasions he emphasized its importance.[3] The themes and problems discussed in the *Doppelvortrag*, in particular the two correlated concepts of *definite theory* and *definite manifold* were later extensively reconsidered in his *Formale und Transzendentale Logik*.[4]

Currently two versions of the *Doppelvortrag* are available, one dating back to 1970, published in *HU XII*,[5] and a more recent version, edited by E. Schuhmann and

---

[1]Husserl, *Das Imaginäre in der Mathematik. I: Zu einem Vortrag in der mathematischen Gesellschaft in Göttingen 1901*; in: *PdA* 430–451; *PoA* 409–452. Henceforth cited as *PdA* App.

[2]Schuhmann and Schuhmann 2001, 87. Henceforth cited as Sch&Sch.

[3]See for example *Ideas* §72; *FTL* [ed. 1929], 85.

[4]*FTL* 78 ff.

[5]*PdA* App. 430–451; *PoA* 409–432.

S. Centrone, *Logic and Philosophy of Mathematics in the Early Husserl*, Synthese Library 345,
DOI 10.1007/978-90-481-3246-1_3,

K. Schuhmann, from 2001.[6] In this second edition the text is reconstructed differently on the basis of a note, unknown in 1970, contained in the *Jahresbericht der Deutschen Mathematiker-Vereinigung*, which states that in the session of December 10, "E. Husserl continues his lecture of November 26. Mainly the concepts of 'definite' system and 'absolutely definite' system are explained. In definite systems, and only in these, the transition through the impossible is admitted." – As for the analysis of what we consider to be, for internal reasons, the main body of the text, we will mainly rely on the first edition. (Our justification for this choice will be presented later on.[7]) But we will refer both to the first and to the second edition whenever we try to find a formal counterpart for certain specific Husserlian concepts discussed in the *Doppelvortrag*. So, for instance, we will rely on the first edition for our analysis of the concepts of *operation system* and *arithmetizability* of a manifold and on the second edition for our analysis of the concept of *mathematical manifold*, since the latter concept is missing from the first edition. We feel free to adopt this procedure, since the editors of the second version admit that as regards the first *Vortrag* there are no relevant problems to recognize the main body of the text but that it is difficult to reconstruct the text of the second *Vortrag*, since the main part of the lecture manuscript is lost: "So this material had to be incorporated into the text in an order for which the editors themselves bear responsibility."[8]

Let us start with some historical remarks about Husserl's personal and scientific relation to Hilbert as well as about the particular context in which Husserl delivered his *Doppelvortrag*; for this will turn out to be very helpful when it comes to articulating the conceptual background of the specific problems Husserl tries to solve in his lectures.

On September 14, 1901 Husserl was appointed extraordinary professor at the Faculty of Philosophy of the University of Göttingen by the Prussian Minister for Religious, Educational and Medical Affairs.[9] His appointment was applauded in the faculty of mathematics: Felix Klein and especially David Hilbert hoped to find in Husserl a colleague who was very much interested in their direction of research.[10] The faculty of philosophy, by contrast, especially two of its two most influential chair holders, the philosopher Julius Baumann and the psychologist George Elain Müller, strongly opposed Husserl's arrival, for they perceived it as a weakening of their own direction of research. At that time Hilbert was trying to establish in Germany an interdisciplinary area of research for mathematicians, logicians and philosophers on the model of the kind of cooperation that was in bloom at Cambridge around Bertrand Russell.[11] Hilbert held Russell's and Whitehead's work in high esteem, and he was "convinced that the combination of mathematics,

[6]Sch&Sch 87–123.

[7]See below §4, footnote 74.

[8]Sch&Sch 88.

[9]Schuhmann 1977, 67.

[10]Cp. Sch&Sch 87; Peckaus 1990, 206 ff.

[11]Peckaus 1990, 210.

philosophy and logic ... should play a greater role in science".[12] He thought it to be very desirable that a least part of the philosophical work in Göttingen be devoted to philosophical questions concerning the axiomatic foundation of the mathematical sciences. Presumably, Hilbert was disappointed by the later transcendental turn of Husserl's philosophy and by Husserl's increasing interest in purely philosophical rather than logico-mathematical problems. When Husserl had left for Freiburg and the discussion about who was to become his successor had begun, Hilbert recommended his former student Leonard Nelson to the faculty of philosophy. Together with other members of his own faculty he signed a document in which it was pointed out that epistemological problems connected with the development of the mathematical sciences did not receive sufficient attention in Göttingen, as witnessed by the fact "that one did not make a man like Husserl stay because one did not recognize his importance (*dass man einen Mann wie Husserl nicht hielt, weil man seine Bedeutung nicht erkannte*)".[13] When Husserl had arrived in Göttingen he had immediately become involved in the activities of the *Mathematische Gesellschaft*.[14] On November 5 he attended (and later wrote down from memory) a lecture by Hilbert given before that Society on the topic of "the closure of axiom-systems (*die Abgeschlossenheit von Axiomensystemen*)". His own *Doppelvortrag* followed in turn, and it is reasonable to assume, both because of the similarity of topics and because of Husserl's particular position in the faculty, that he tried to reconsider and to answer questions that Hilbert had posed on that very occasion.

The text of 1901 that we shall analyze is entitled by Husserl himself "*Das Imaginäre*".[15] It has three parts. In the *Einleitung* Husserl presents the problem as a fundamental question (*Grundfrage*) which concerns the mathematical *method*, underlining its significance for both mathematics and philosophy. In the second part, *Theorien über das Imaginäre*, he briefly examines five proposals for a solution put forward by contemporary mathematicians, and refutes them all. In the course of his analysis of these alternative theories Husserl appeals to the idea of a concrete realization of an *arithmetica universalis*. In the third part, *Der Durchgang durch das Imaginäre*, he presents his own attempt to solve the problem.

His main aims are

(a) to elaborate the structure of an *arithmetica universalis*, or *general theory of arithmetic*, in the direction of an even more general *theory of deductive theories*
(b) to show how the solution of the 'problem of the imaginary' is closely connected with a specific property that only some formal theories have, – the property of 'definiteness'.

[12]Reid 1970, 144.

[13]Hilbert's et al. "*Minoritätsgutachten*" of 1917 against the majority report of the Faculty. Quoted after Peckaus 1990, 210.

[14]Cp. Sch&Sch 87.

[15]*PdA* App. 431, fn.; (not translated in *PoA*).

The following pages contain

(i) an analytic reading of Husserl's text, with particular attention to these two indicated topics
(ii) a mathematical interpretation of the Husserlian concepts of 'universal arithmetic', 'definite axiom-system' and 'definite manifold'
(iii) a discussion of divergent interpretations of Husserl's two notions of definiteness
(iv) an attempt at a formal reconstruction of Husserl's theory of manifold, also in the light of some further clarification on this same topic given by Husserl in his later *Formal and Transcendental Logic*.

3.2 The *Einleitung*

The *Einleitung* outlines the basic elements of the issue and opens with the following statement: "The theme which I wish to deal with in this lecture concerns a fundamental question (*Grundfrage*) of the mathematical *method* and belongs as such to that difficult field in which mathematicians and philosophers are interested to the same degree, even if not entirely in the same sense".[16] Husserl is referring here to the *fundamental question* put to mathematics, with increasing frequency from 1850 onwards, concerning the proliferation of "new mathematical concepts", that is concepts that seem to lack any content but nevertheless are "useful" in the practical process of calculating.[17] The general thesis that Husserl presents in the *Doppelvortrag* can be formulated as follows: if a new concept in mathematics is useful, then the need arises both for mathematics and for philosophical reflection on mathematics to investigate the principles that are at its basis, in order to confer to it a status analogous to that of already accepted arithmetical concepts. In this case the *use* of the new concepts is *legitimate* or *justified*, which means that it does not lead to contradictions. The *Doppelvortrag* hence sketches the general guidelines along which a rigorous (mathematical) justification of the use of these new concepts can be brought about.

Incidentally, the question was not at all alien to Dedekind who wrote in *Was sind und was sollen die Zahlen?*:

> The greatest and most fruitful advances in mathematics and other sciences have invariably been made by the creation and the introduction of new concepts, rendered necessary by the frequent recurrence of complex phenomena which could be controlled by the old notions only with difficulty.[18]

---

[16]*PdA* App. 430; *PoA* 409. Cp. Hartimo 2007, 298 ff.

[17]Cp. Sieg 2002.

[18]Dedekind 1888, VI. Already in 1854 Dedekind had delivered a lecture on this topic before the Mathematical Society in Göttingen on the occasion of his admission as *Privatdozent* (loc. cit.)

Now Hilbert's identification of the mathematical existence of a concept with the consistency of the system of axioms involving that concept is motivated by the very same problem (the proliferation of new and useful mathematical concepts).[19] Hilbert confronted this problem in the initial phase of his reflections on foundations when he aimed at an axiomatic foundation not only of mathematics but also of physics and other sciences through the formal-axiomatic method.[20] This phase spans from 1898 to ca. 1901, and these are just the years in which Husserl's Göttingen period began. In the famous conference on "*Mathematical Problems*" held in Paris in 1900 during the second International Congress of Mathematicians, Hilbert posed questions about the *meaning* of mathematical problems for the development of mathematics and about the *sources* from which mathematics derives its problems.[21] Some of his remarks turn out to be very useful to sketch the context and set of problems from which Husserl's *Doppelvortrag* arises:

> Just as any other human undertaking pursues goals, so mathematical research needs problems... It is difficult, and often impossible, to estimate the value of a problem in advance, for in the end what is decisive is what science gains by pursuing that problem. Nevertheless we can ask whether there are general characteristics that mark a good mathematical problem... Surely the first and oldest problems in every branch of mathematics spring from experience and are called forth by the world of external phenomena. Even the rules for *calculating* with integers ... have been discovered in this way... But in the progressive development of a branch of mathematics the human mind, encouraged by the success of its solutions, becomes aware of its autonomy. It creates by itself new and fruitful problems – often without any recognizable external stimulus, just by logically combining, by generalizing and particularizing, by separating and collecting concepts in the most felicitous way, and thus the human mind itself steps into the foreground as the real questioning subject (*der eigentliche Frager*)... But then, whilst the creative power of pure thinking is at work, the external world again comes into play, actual phenomena force upon us new questions and open up new branches of mathematics, and, while trying to incorporate new fields of knowledge into the realm of pure thinking, we often find the answer to ancient and unsolved problems, thereby improving old theories in the best possible way. The numerous and surprising analogies and the apparently pre-established harmony between the questions, methods and concepts of the various branches of knowledge which the mathematician so often perceives all have their origin, it seems to me, in this ever recurring and ever changing interplay between thought and experience.[22]

Hilbert's main problem at that time can be recast as follows: "if a new concept is useful in mathematics, how can we affirm that it mathematically exists?"[23] At this stage of his foundational research his answer is this: "if one succeeds in proving that

---

[19]This proposal was anticipated by Georg Cantor, who in 1883 wrote "mathematics is entirely free in its development and its concepts are restricted only by the necessity of being non-contradictory and coordinated to concepts previously introduced by precise definitions" (Cantor 1883, 563–64; transl. in Kline 1972, 1031.

[20]Abrusci 1978, 19 ff. See also Abrusci 1981, Corry 2004.

[21]Cp. Reid 1970, 70–71.

[22]Hilbert 1900b.

[23]Abrusci 1978, 21.

the constitutive properties of a concept cannot ever lead to a contradiction using a finite number of logical inferences,"[24] then the mathematical existence of that concept is proven. In 1900 Hilbert sought to establish the consistency of the axiom-system for the real numbers: "such a proof was to ensure the existence of the set or, in Cantor's terminology, of the consistent multiplicity of the real ... numbers."[25]

It should be clear at the outset that Husserl's problem in the *Doppelvortrag* is *not* that of finding a proof of consistency for the system of axioms involving some new mathematical concept. Rather, Husserl simply assumes its consistency as a hypothesis. Nevertheless, the *Doppelvortrag* can be read as Husserl's own contribution to the solution of another problem related with Hilbert's concerns at that time. He will try to answer the following question: under which conditions can the consistent system of natural numbers be stepwise expanded to other numerical systems, up to the system of the real numbers?

Husserl presents the problem of the imaginary as a problem that arises in the context of the evolution of mathematics from a science of numbers and quantities into a theory of arbitrary abstract structures. He is thinking of the radical transformation that mathematics underwent during the nineteenth century, which Howard Stein defines as a transformation "so profound that it would not be inappropriate to call it a rebirth of the subject".[26] The transformation consists in a change in the very way of conceiving mathematics, from a science of systems of determinate entities, to a study of multiply exemplifiable abstract structures.[27]

As Casari rightly stressed, Husserl "really had understood the fundamental aspects of the development of mathematics in his time: its going towards formalization, algebraization ... ",[28] and he was "fully aware of the fact of having had this idea before anyone else; some years later, in fact, he will say something like: 'now everyone is talking about formalization, but then I had been the only one to see this thing, to build it up laboriously with theoretical and historical studies'".[29] Husserl explicitly refers to Leibniz' idea of abstract mathematics as *mathesis universalis*, as a *theory of theories* that is capable of determining the *general form* of all formal-mathematical or deductive disciplines.

> Originally limited to the field of numbers and quantities, mathematics has grown far beyond that field. It has increasingly approximated to the goal that *Leibniz* had already clearly conceived, namely, the goal of being a pure theory of theories (*Theorienlehre*), free of all

---

[24]Hilbert 1900b.

[25]Sieg 2002, 365.

[26]stein H 1988, 238–259; quoted after Sieg 2002, 365. Cp. Tieszen 1995: "Husserl ... was witness to advances in formalization, generalization, and abstraction that were unprecedented in the history of mathematics" (50).

[27]Cp. Cavaillès 1938: "Les mathématiques réelles initiales ne sont plus qu'un cas particulier situé au sein des mathématiques nouvelles, expliqués par elles" (54).

[28]Casari 2000; cp. Tieszen 2005, 9.

[29]Casari 2000; cp. also Tieszen 2004, 29.

> special fields of knowledge and insofar formal. Mathematics in the highest and most inclusive sense is the science of theoretical systems in general, in abstraction from that which is theorized in the given theories of the various sciences.[30]

But Husserl is also always interested in studying the ontological correlate of formal theories, that is to say, he is always aware of the fact that to formal theories correspond "possible fields of experience" that are axiomatized by those theories. Hence abstract mathematics, as *mathesis universalis*, is also to be considered as the most general theory of structures (each constructed with its own axiom-system), the task of which is to create step-by-step suitable instruments for the interpretation of important parts of the world of experience.[31]

The first step in the construction of a formal theory consists in *formalization*. To 'formalize a theory' means *to abstract from the matter*, from the particular concrete "field of experience" that the theory describes and to consider its form. This is done by substituting "object variables" for the names of "materially determinate objects". Thus, for example, we substitute the letters a, b, c, ... etc. for the designations of natural numbers. The properties of the objects are now specified by the axioms of the theory. The formalization allows the unification of fields of experience that appear to be vastly different: once the theories are formalized, it becomes obvious that these fields are axiomatized by the same theory.[32] In this way a "generalization (*Verallgemeinerung*)" is performed: the concrete theory is now an instance of a class of theories that all have the same form or, as Husserl puts it, an instance of a certain "theory form (*Theorienform*)". It is only later, in *Ideas*, that Husserl states explicitly that the transition from a concrete theory to its form (*Formalisierung*) and the transition from a certain formal theory to another more general theory (*Verallgemeinerung*) are not the same thing: indeed, in the first case we have a transition from the material to the formal, in the second a relation among theory forms. In the *Doppelvortrag* under examination, however, Husserl does not confuse the two concepts, but only their names.[33]

At the time of this lecture Husserl conceives of a formal theory, in the narrow sense of the word, as a collection of axioms that are *purely formal*, *mutually consistent* and *independent* and, in a broader sense, as also including all propositions derivable from the axioms "in a purely logical way (*rein logisch*)", that is the *theorems* of the theory. The related *field* (*Objektgebiet*) is in turn considered as "a field of objects in general", determined only by the fact that it 'falls under' certain

---

[30] *PdA* App. 430; *PoA* 409–410. In the *Prolegomena* we read: "The evident possibility of generalizing (transforming) formal arithmetic, so that, without essential alteration of its theoretical character and methods of calculation, it could be taken beyond the field of quantity, made me see that quantity did not belong to the most universal essence of the mathematics or the 'formal', or to the method of calculation which has its roots in this essence" (*PR* VI, *PRe* 41–42).

[31] This point has also been strongly emphasized by Hilbert, his "favorite example being that [of the application of] of the Euclidean axioms of linear order and congruence to the genetic variations in *Drosophila* flies produced by cross-breeding" (Webb 1980, 81). Cp. also Casari 2000.

[32] *PdA* App. 431; *PoA* 410.

[33] *Ideas* §13; cp. *FTL*, 81. See also Tieszen 2004, 28.

axioms of this or that form. "We call an object field thus defined a [...] formally defined manifold."[34] The axiom-system that characterizes a formal manifold hence determines only the relations that occur among the elements of the manifold, leaving the nature of the elements undetermined.

The requirement of *formalizing* a concrete theory, of transforming it into a *formal axiomatic theory*, is articulated by Hilbert in what was referred to above as the "first phase" (1898–1901) of his foundational research. At that time, his use of the formal axiomatic method is characterized by methodological features like (i) the tacit assumption that every logical consequence of the axioms is derivable from them in a finite number of logical steps; (ii) the fact of not making the language and logic of a formal theory explicit; (iii) the examination of the derivability or non-derivability of given theorems from certain groups of axioms. He also requires a formal theory to satisfy a number of additional conditions, including (1) the finiteness of the number of axioms, (2) the independence of the axioms (i.e. the non-derivability of each of the axioms posited for a certain theory from the remaining ones), and (3) the reduction of the axioms to the least possible number. All these conditions are also implied in Husserl's conception of formalizing a theory in the *Doppelvortrag* of 1901.[35]

Furthermore (in this same phase of his foundational research), Hilbert thinks that it is possible to reduce the axiomatic foundation of all of mathematics to that of the arithmetic of real numbers and set theory. For the axiomatic foundation of the theory he requires not only its formalization, but also a proof of the consistency of the axioms of the theory. This latter constraint, essential for Hilbertian foundationalism, constitutes an important difference from Husserl's conception of formalism at the time, as Husserl, in essence, thinks that once we have stated the axioms for a certain formal theory, it is "reasonable" to assume that they are non-contradictory.[36]

Already at this stage in the development of Husserl's thought the idea is present that the forms of the theories obtained by formalization can be systematically classified and related and that they can be put in connection with classes of theories with a different form. On the basis of these interrelations significant theoretical conclusions can be drawn, especially with respect to the fact that formal theories as well as the structures corresponding to them can be *generalized* or *specialized*, can undergo 'expansions' and 'restrictions'.[37]

> The theory forms defined by such abstraction can, then, be set into relation to one another; they can be systematically classified; one can broaden or narrow such forms; one can bring a certain previously given theory form into systematic interconnection with other forms of exactly defined classes and draw important conclusions concerning their interrelationship.[38]

---

[34] *PdA* App. 431; *PoA* 410.

[35] Also cp. Tieszen 2004, 29 and Tieszen 2005, 9.

[36] For an opposite view on this point see Ortiz Hill 1997b, 153.

[37] As Hartimo 2007 puts it (though with reference to another work of Husserl's): "Husserl is occupied by some kind of a structural relationship between two different domains" (296).

[38] *PdA* App. 431; *PoA* 410.

The geometry of our physical space, for instance, is a concrete theory that yields, when formalized, the form of theory that we call a 'theory of the three-dimensional Euclidean manifold'.[39] In maintaining that through formalization theories are transformed into mathematical constructs and hence become objects of mathematical study[40] Husserl anticipates a point Hilbert made a few years later. Hilbert expresses this idea for the first time – but only with respect to formal proofs, not to theories – in his conference paper *Über die Grundlagen der Logik und der Arithmetik* (1904):[41] "one has to consider the proof itself as a mathematical construction (*Gebilde*)" and, hence, as an object of mathematical research. Moreover, the distinction is present in Husserl between a theory as a formal system and a theory as a manifold, that is the structure underlying a theory conceived as deductive system. Consequently, he studies purely formal relations of generalization and specialization not only among theories but also among manifolds,[42] and he poses the problem of identifying relevant correspondences between certain properties of the formal theories and certain properties of the manifolds.

The problem of the imaginary is introduced by Husserl in these terms:

> Mathematics is thus, according to its highest-level conception, a theory of theories, the most general science of possible deductive systems in general. This generalization, through which the sphere of the old objects of mathematical investigation – the cardinal numbers, ordinal numbers, the scalar and vectorial magnitudes and the like – is entirely transgressed, is the source of unsolved methodological problems.[43]

The project of elaborating the structure of a formal mathematics (*formal arithmetic* or *general* or *universal* arithmetic) that would be a general theory of deductive systems is connected with reflections on the specific problem of the 'imaginary in mathematics'. It has to be kept in mind that Husserl uses the term 'imaginary (numbers)' in a very broad sense, as a collective name that encompasses negative numbers, rationals, irrationals, complexes, that is all numbers except the whole positive numbers.[44] Husserl observes that the unchecked use of the symbolism in algebra unleashed from its objective reference – a consequence of what he calls the "tendency toward formalization" in algebra[45] – is at the basis of

---

[39]Loc. cit. Cp also Hartimo 2007, 299.

[40]Cp. *FTL* 79.

[41]Published as Hilbert 1905.

[42]Cp. *FTL* 80. For a formal characterization of Husserl's concepts of generalization and specialization see below §§5 and 10.

[43]*PdA* App. 431; *PoA* 411.

[44]*PdA* App. 432–433; *PoA* 412.

[45]*PdA* App. 432; *PoA* 412. Cp. Webb 1980: "The history of algebra has indeed gravitated to this formalistic principle. The desire to solve all algebraic equations leads to the notion of an algebraic closed field constructed by 'formally adjoining' new elements to a given field to serve as solutions to equations which had none over it. If a polynomial $P(x)$ has no roots in a field F, this does not imply the inconsistency of the claim that $(\exists X)(P(X) = 0)$generally, hence it should be satisfiable in a suitable extension F' arising from F by the formal adjunction of new 'numbers' to F' (86).

> forms of operation which were arithmetically meaningless, but which manifested the remarkable character that they could nevertheless be utilized in calculations. For it turned out that if the calculation was mechanically executed according to the rules of operation, as if everything were meaningful, then, at least in a broad range of cases, every result of calculation free of the imaginaries could be claimed as correct, as one could empirically establish by means of direct verification.[46]

Faced with this difficulty, institutional mathematical science reacts by increasingly perfecting the techniques of calculation, without any concern for the *difficulties in principle* that exist *in the application of the symbolism to different concrete numerical fields.* The problem of the imaginary, this is the point on which Husserl insists, is a problem that concerns the *methodology* of mathematics, and it is rooted in the fact that reflection on the symbolic or formal aspect has not been generated by a genuine theoretical interest, but by a practical interest directed at the development of arithmetical algorithms and satisfied by finding new rules of calculation to solve various concrete mathematical problems.[47] Husserl had already made this point in the *Philosophy of Arithmetic*, where 'the general arithmetic of cardinal numbers (*allgemeine Arithmetik der Anzahl*)' was conceived of as a 'general theory of operations', the fundamental task of which was to find the greatest possible number of procedures to make the methods of calculation increasingly fast and efficient.

Up to this point Husserl's argument in the *Einleitung* of the *Doppelvortrag* can be summarized as follows:

(a) In the first form of pure mathematics, algebra, the rules of calculation that are valid for the arithmetic of finite cardinal numbers are abstracted from the conceptual field that they are meant to interpret – the numerical field – and considered independently from the original domain. The arithmetical algorithm that is now a complex of symbols regulated by formal rules becomes the object of mathematical research.

(b) Like arithmetic, all concrete theories (of exact sciences such as geometry) can be transformed into formal axiomatic theories and, once formalized, they can be expanded in a purely formal way.

(c) "Expanding the rules of calculation that are valid for the theory of finite cardinal numbers in a purely formal way" means: removing restrictions on arithmetical operations that are valid for whole positive numbers, and, Husserl observes, this is done *without considering the correlative modification of the concrete numerical field.* This situation is the source of the difficulties that concern the arithmetical method, the most important of which is related to the application, in calculations, of 'meaningless forms' that nevertheless yield correct results. From a more general point of view, the problem of the imaginary consists essentially in the lack of understanding of the relations that obtain

---

[46] *PdA* App. 432; *PoA* 412.

[47] Cp. Cavaillès 1938: "La fécondité est l'instance devant laquelle tout réfus au nom de l'évidence s'avère préjugé" (54).

between *general algorithm* and *real mathematics*, between *forms of theories* and *concrete theories*.

Further explanations of (a)–(c) in the *Einleitung* show how in Husserl's conception of axiomatization two ideas co-exist: (1) the objects of a given field of experience constitute a system of things whose reciprocal relations are regulated by axioms which constitute the theory of that field of experience, and (2) if the axiomatization is to be meaningful there has to be a *content* which the axioms express. The field of a science, Husserl writes, is delimited by a general concept (*allgemeiner Begriff*) and constituted by the objects that fall under it. For instance, the field of cardinal numbers is determined by the general concept of '*Anzahl*' and is constituted by the objects that fall under it, the *Anzahlen*, understood as the 'particularizations of the general concept'. The relations and connections that are possible between these objects flow from the very nature of these objects and find a *formal expression* in an axiom-system that constitutes the theory of the field.[48] For this theory, that is *on the basis of the specific nature of the objects of the field*, certain '*compositions*', certain '*operational complexes*' ('*imaginary forms*') do not make sense. Nevertheless, when used in calculations, they yield correct results.

> Suppose a field of objects given in which, through the peculiar nature of the objects, forms of combination and relationship are determined that are expressed in a certain axiom-system A. On the basis of this system, and thus on the basis of the particular nature of the objects certain forms of combination have no real signification (*reale Bedeutung*), i.e. they are absurd forms of combination.[49]

At this point the fundamental question for Husserl is this: *can 'countersense (das Widersinnige)' justifiably be used in the calculus*? With what justification can rational deductive thought, whose scientific expression is a formal-mathematical science, admit 'countersense', and accept the results based on it? *If 'countersense' has to be admitted in the calculus, then how can that be theoretically justified?*

3.3 Universal Arithmetic

> *Or, il faut élargir le calcul. Puisque les objets ne déterminent pas, dans une intuition désormais impossible, leur mode d'emploi spécifique, ils ne seront que des supports pour un certain traitement: la certitude de leur connaissance – et son contenu – ne peuvent provenir que de l'exact enchaînement des opérations qui leur conviennent. Donc, autant de calculs que de théories mathématiques et un calcul général qui les subsume tous et qui ne peut être qu'une théorie formelle des opérations.*[50]

---

[48] Cp. Chapter 2, §4 of this work.

[49] *PdA* App. 433; *PoA* 412.

[50] Cavaillès 1938, 56.

By 'universal arithmetic' Husserl means a system of axioms (generally conceived of as equations) which govern the behaviour of arithmetical operations and are valid in all numerical systems (the wholes, the rationals, etc.). Different fields of knowledge, each determined by a given general concept and constituted by different objects, *can be interpreted by the same algorithm*, in other words, *the objects of these fields can stand in the same relations and can be axiomatized by the same theory*. The concepts of natural number, whole number, rational number etc. determine different numerical systems, for the genus of the objects by which they are constituted changes. The relations among the natural numbers, the "rules of calculation" for the elementary operations on the naturals, can be expressed in an 'axiom-system' that Husserl calls the 'general arithmetic of the *Anzahl*'. The situation is analogous for whole numbers, rational numbers, etc.: each numerical system will be axiomatized by a different arithmetic.

Numerical systems have the peculiarity that the objects of their fields can be ordered by a total order relation (obviously, with the exception of that of the *complex* numbers), and in all these systems certain operations (+, −, ...) are given. For Husserl, universal arithmetic is the *part* that all these theories have *in common*. In other words, the different arithmetics are considered as specific instances of universal arithmetic; the latter represents the general form of all these theories and stands in a relation of genus to species with respect to them: it is the system of rules of calculation that is valid in all numerical systems.

Among the rules of calculation Husserl distinguishes

(a) general forms of operations (*allgemeine Operationsformen*)
(b) specific forms of operations (*Operationsformen der Besonderung*), that is operations that are specific for a particular numerical system.

The general forms of operations, in addition to the axioms and the rules they have to comply with, constitute universal arithmetic, while the specific forms of operations determine the different arithmetics in their specificity. The passage from one arithmetic to another, for instance, from that of the natural numbers to that of whole numbers, consists in expanding the arithmetic of the natural numbers by determining some of the specific forms that had been left open by the general axiom-system, that is by universal arithmetic.

> The old axioms give a determinate sense to the general operations and to certain special forms of operation. What is not defined, that is excluded in this narrower field. The new axioms retain all of these axioms, but give sense to special operational formations which previously were not defined and were previously left open.[51]

So the system of natural numbers, the system of whole numbers, etc. are conceptual fields that are 'founded' by different concepts. But, they all fall under the *universal arithmetic*, under a common system of rules of calculation. The specificity of the single arithmetics is given by specific forms of operations, rules of calculation that are valid only in one or more (but not in all) numerical systems.

---

[51] *PdA* App. 444; *PoA* 431–432.

The idea is that specific forms of operations can be admitted in one field without making sense in another field: in certain fields they may be 'imaginary forms'.

> Let us consider, for example, the axiom-system of the whole numbers, positive and negative. Then $x^2 = -a$, $x = \pm\sqrt{-a}$ certainly has a sense. For square is defined, and –a, and = also. But "in the field" there exists no $\sqrt{-a}$. The equation is false in the field, since such an equation cannot hold at all in the field. Therefore I cannot pose the problem: "A certain magnitude [x satisfies] $x^2 = a$. Which magnitude is that?"[52]

As we shall see in the course of our analysis, Husserl's investigation of the conditions under which the gradual extension of the system of natural numbers is possible can clearly be regarded as an application of what Hilbert in his paper *Über den Zahlbegriff* calls the '*genetic method*',[53] and of the strategy at the basis of Dedekind's gradual extension of the number-concept by reduction of the new numbers (negative, rational, irrational and complex numbers) to the naturals. (As we shall soon see, there is however a conceptual divergence between Husserl and Dedekind as to what is to be expanded.) Hilbert's argument in the paper mentioned above exactly models the idea of a stepwise expansion of the consistent multiplicity of natural numbers that Husserl tries to realize with his account of the relation between universal arithmetic and specific arithmetics. Hilbert says:

> Let us first pay attention to the way in which the concept of a number is introduced. Starting from the concept of the number 1, one normally thinks of the other positive integers 2, 3, 4, ... as arising from the procedure of counting and of their rules of calculation as developed in this process; then one arrives at the negative numbers by requiring the general executability of subtraction; then the rationals are defined e.g. as pairs of numbers, and thus every linear function has a zero; finally one defines the reals as sections or fundamental successions and consequently, one obtains the result that each whole indefinite rational function has a zero and that the same holds for each indefinite continual function. We can call this method of introduction of the concept of a number *genetical method* because the most general concept of real number is obtained by way of successive expansions of the concept of natural number. [54]

3.4 Theories of the Imaginary

To sustain his conception of universal arithmetic, Husserl presents and criticizes five different "theories of the imaginary." Regarding the first two, we will restrict our discussion to the few hints that he provides about them. With respect to the third and the fourth theory, the most significant aspect that emerges from his criticism consists in the elaboration of the conception of universal arithmetic that we have just outlined. But the fifth theory, because of the extreme interest of the logical reflections

---

[52]Sch&Sch 111–112; *PoA* 438–439.

[53]Hilbert 1900a, 180–181.

[54]Loc. cit.

that emerge from its exposition and refutation,[55] deserves a separate and detailed analysis.

The first two theories are those of Bain and Baumann, resp. of Boole. The *first* states that "the imaginary is vindicated empirically, through induction", the *second* that "the imaginary is directly evident a priori".[56] For Husserl, the unfeasibility of these theories is so obvious that they are presented without discussion.

In his sketch of the *third* theory Husserl clearly uses Dedekind's essay *Stetigkeit und irrationale Zahlen*[57] as his point of reference. In this essay Dedekind shows that the "reduction" of the *reals* to the *rationals* through the creation of an *arithmetized* model of the geometrical line is possible. His argument runs as follows: Because numbers are constructed by a *free creative procedure*, we can progressively expand the originally defined numerical system, that of natural numbers, in such a way that all inverse operations become effectively executable. This expansion must be effected by *definitions*: new numbers are introduced by definitions, insuring that the rules of calculation for these new numbers wherever possible follow those that are valid for the numbers of the system with which we started and do not give rise to contradictions in the whole system.

The idea of a series of successive expansions of the concept of natural number, that is the creation of negative numbers, of rationals, of irrationals, and, finally, of complex numbers by reducing the laws of calculation that are valid for these numbers to those that are valid for the natural numbers is unfeasible by Husserl's lights – at least in the way in which it is proposed by Dedekind. Husserl's critique is not focused on a logical difficulty in Dedekind's theory (as it will be in the case of the "theory of permanence"), but rather on a more philosophical problem: the formal procedures by which the expansion of the natural numerical field is obtained are correct, but Dedekind's *conceptual* presuppositions concerning the foundation of that expansion are not acceptable. The core of Husserl's argument is that *one cannot expand the concept of natural number* (*Anzahl*). A natural number, by definition, serves as an answer to a 'how many' question. Something that is not a cardinal number cannot serve as an answer to such a question. Hence it cannot properly be called a number. The field of natural numbers is univocally determined by a general concept and by the possible operations within that field, which are also based on that general concept.

> I cannot, without absurdity, arbitrarily expand the sphere of the concept of *Anzahl* on the basis of creative definitions, for this very concept imposes limits on me... Once a word – e.g., the word "*Anzahl*" – is confined to a given field of objects, one that clearly presents itself as possible, then I cannot decree by some sort of arbitrary stipulation that the field is to admit an expansion by means of new objects. It would be as if in geometry one would decree: There are round squares, if not in the plane, then in a higher dimension of space.[58]

---

[55]At this point Hartimo 2007, 300 f. commits a serious mistake. Husserl does clearly *not* adopt the fifth theory as she wrongly suggests.

[56]*PdA* App. 434; *PoA* 413.

[57]Dedekind 1872.

[58]*PdA* App. 435; *PoA* 414–415.

Hence, for Husserl, it is not possible to enlarge the system of natural numbers by new objects (e.g., whole negative numbers). What we can do, and this is the crucial point of Husserl's critique, is to assume (with reference to the example of the negatives) a *new, purely formal concept*, that of whole number. This concept does not answer the question 'how many?', and it is introduced by a definition in a purely formal way, *via* "the formal system of the definitions and operations valid for the *Anzahlen*". In other words, we use the same formal system to interpret a different conceptual field, one that is determined by the concept 'whole number'. Only in relation to this new field can the system of definitions and operations be expanded by new definitions for the specific forms of operations that are valid, in our example, for the whole negative numbers. The new system of operations obtained in this way partially coincides with the original system (in the sense that we find in it a subsystem that "behaves like" the natural numbers), and at the same time it is broader, as it "contains more basic elements and more axioms."

On the same basis Husserl also rejects a fourth theory that is meant to cover all positions that deduce the legitimacy of imaginary magnitudes from the real existence of different kinds of magnitudes: distances, temporal magnitudes, etc.[59] So the factual existence of different magnitudes is supposed to be the origin of different numerical concepts. For Husserl, however, empirical proofs are not sustainable in theoretical contexts. Furthermore, such a conception is founded on an insufficient analysis of the relation between the different 'kinds of number (*Zahlarten*)' and the related arithmetics.

So far Husserl's result is this: different numerical concepts 'ground' different arithmetics, and these "do not have parts in common; rather, they have wholly different spheres, but an *analogous structure; they have partially the same forms of operations*, although different concepts of operation."[60]

The *fifth* theory refuted by Husserl is the "principle of the permanence of the formal laws" of Hermann Hankel which essentially consists in the following requirement: when in mathematics we want to expand a concept beyond its original definition (in particular, the concept of number), we must proceed in such a way as to *preserve*, as far as possible, the "old" rules of calculation.[61] In this context

[59]Cp. Dedekind's analogous expulsion of the notions or intuitions of space and time (Dedekind 1888, III) and of measurable quantities (1872, 9–10) from arithmetic.

[60]*PdA* App. 438; *PoA* 416 (my emphasis).

[61]The main contributions of the mathematician (and historian of mathematics) Hermann Hankel (1839–1873) concern the theory of functions and of complex and hypercomplex numbers. The 'principle of permanence of formal laws' is formulated in his *Theorie der complexen Zahlensysteme* Hankel 1867, where the system of complex numbers, that of Hamiltons quaternions, and additionally some of H. Grassmann's algebraic systems are presented in great detail. The 'principle' is a revision and a precisification of the 'principle of permanence of equivalent forms' introduced by the algebraist G. Peacock (1791–1858) to warrant the meaningfulness of the passage from arithmetical algebra to symbolical algebra. Essentially, this principle consisted in the requirement that the laws of arithmetical algebra should also be laws of symbolical algebra: if the general rules for arithmetical operations are adapted to the corresponding operations of symbolical algebra, then we will have absolute identity of results for the common part of the

Husserl discusses the concept of a 'definite (*definit*)' axiom-system, and he makes essential use of it when it comes to solving the problem of a theoretical justification of the use of the imaginary in mathematics. To obtain what he calls 'a passage through the imaginary (*ein Durchgang durch das Imaginäre*)', Husserl accepts some of the results that the theory of permanence arrives at, although he does not regard them as valid in general but only for *definite* axiom-systems.

In a passage of *Formal and Transcendental Logic* Husserl explicitly refers to these pages of the *Doppelvortrag*, and he gives a very clear and precise summary of the problem as well as of the particular solution that he advanced there:

> The concept of the definite manifold served me originally for a different purpose, namely to clarify the logical sense of the computational transition through the "imaginary" and, in connection with that, to bring out the sound core of Hermann Hankel's renowned, but logically unsubstantiated and unclear, "principle of the permanence of the formal laws." My questions were: Under what conditions can one operate freely, in a formally defined deductive system (a formally defined "manifold"), with concepts that are imaginary – according to the definition of the system? When can one be sure that deductions that involve such an operation, but yield propositions free from the imaginary, are indeed "correct"- that is to say, *correct consequences of the defining forms of axioms*? How far it is possible to "extend" a "manifold", a well-defined deductive system, into a new one that contains the old one as a "part"? The answer is as follows: If the systems are "definite", then calculating with imaginary concepts can never lead to contradictions.[62]

The argument that Husserl develops in the *Doppelvortrag* begins with an *abstract* reconstruction of Hankel's theory of permanence and its logical grounds. Let G be a certain given concrete field, for example, that of the natural numbers. Let A_G be the set of axioms relative to G, that is the *formal* field obtained by formalization from G. Finally, let F_G be the set of propositions logically derived from (i.e. the set of the logical consequences, *Folgen*, of) A_G.

> Given this formalization, to each proposition determined by the axioms (*Grundsätze*) of the real field [G] there corresponds a proposition in the formal field, and conversely. The formal field will have the same limitations (*Schranken*) as the real one, limitations that are already fixed (*präformiert*) in the axioms.[63]

Because Husserl is thinking of theories of an algebraic kind, that is, theories essentially characterized by equations, by 'restrictions' of the formal field we must understand restrictions imposed by the axioms A_G to the field of definition, that is to the executability of certain operations. (If G is the field of natural numbers, for example, then subtraction is not total, it is defined only under certain conditions imposed on the minuend and subtrahend.)

---

disciplines; in other words, symbolical algebra will be a "conservative" (and hence consistent) expansion. It is exactly the *question of conservativity* that Husserl will address, as we will see. For a detailed account of Hankel's principle of permanence see Hartimo 2007, 285 ff.

[62] *FTL*, 85.

[63] *PdA* App. 439; *PoA* 418–419.

In Husserl's reconstruction, the theory of permanence considers at this point an expansion of A_G, let us say A_Γ, with $A_\Gamma = A_G + A'$. Consequently it holds that

$$A_G \subseteq A_\Gamma \textit{ and } F_\Gamma = F(A_G + A').$$

The new axioms A' are such that they exempt the operations from certain restrictions (in the sense explained above). For Husserl this is equivalent to admitting 'imaginary entities'. If, for instance, in the new axioms A' we impose the condition that subtraction is to be definite everywhere, the imaginary entities that we admit are the negative numbers. "Now let us conceive the formal field as expanded in such a way that, as far as is in general possible, it no longer has these limitations."[64]

The theory of permanence at this point states that *if the expansion* (A_Γ) is *consistent then it is also*, as one would put it nowadays, *conservative* over A_G. In Husserl's own words:

> We rise, according to the principle of permanence, above the particular field,[65] pass over into the sphere of the formal, and there we can freely operate with -1. Now the algorithm of the formal operation is indeed broader than the algorithm of the narrower operations... But *if the formal arithmetic is internally consistent*, then the broader operation can exhibit no contradiction with the narrower one. Therefore *what I have formally deduced in such a way that it contains only signs of the narrower field must also be true of the narrower field.*[66]

For Husserl this implication is not obvious at all. More generally, the entire argument hides two distinct problems:

(1) Under what conditions is A_Γ, the expansion of A_G, *consistent*?
(2) Under what conditions is the theory A_Γ *conservative* over A_G?

"Under what conditions are the propositions (*Sätze*) that are free of absurdity (*Widersinn*) also actually valid?"[67] In other words, under which conditions are the expressions "of the old language" (that of G, as we would say today) which are provable starting from A_Γ also provable from A_G, and, hence, true in the concrete field G?

Husserl's fundamental observation is that, *although conservativity* over a consistent theory *implies consistency*, the converse does not hold; in particular, the conclusion that the extended theory is conservative does not follow from the premise that it is consistent, as the theory of permanence seems to maintain. Whether conservativity really obtains has to be proved separately for each theory.

It is natural to stipulate that the 'expanded' theory (i.e. the one that expands) A_Γ is consistent: "An obvious presupposition of the expansion is that the new

---

[64] *PdA* App. 439; *PoA* 419.

[65] Here Husserl refers to the expansion of the system of *real* numbers constituted by the system of *complex* numbers.

[66] *PdA* App. 438; *PoA* 418 (my emphasis).

[67] Loc. cit.

axiom-system be internally consistent. For from what is inconsistent one can obviously prove everything."[68] Under this hypothesis – Husserl argues *impeccably* – given that A_G is included in A_Γ it follows, for any formula α in the language of G, that if α can be proven from A_Γ then surely (because of the consistency of A_Γ) α *is not incompatible with* A_G; that is the negation of α is not provable from A_G. "It certainly is correct that no derived proposition ... can contain an inconsistency, that it can conflict (*streiten*) neither with the expanded axioms nor with the original and restricted axioms."[69]

Now from the fact that α is not in contradiction with A_G (i.e. $\neg\alpha$ is not provable from A_G) the theory of permanence concludes that A_G proves α. In other words, according to this theory, if α does not contain imaginary ("impossible") constructs and follows from A_Γ then – putting it, with Husserl, in semantic terms – it is *true* of the concrete field G:

> If the new system is consistent (*verträglich*) and includes the old one in itself, then in the entire range of deduction no inconsistency can occur. Thus, a proposition which is derived in such a way that it contains none of the "impossible" forms of operation, cannot possibly include an inconsistency, and thus it is true.[70]

However, for Husserl this presupposes a "conceptual jump" that is possible, as will turn out, *only under specific conditions imposed on the initial axiom-system* A_G.

> But how do we know that what is free of contradiction is also true; or, as it must be expressed here, *how do we know of a proposition that exclusively contains concepts which occur in the narrower field*[71] *and are there defined, and which does not conflict with the axioms of the narrower field, that such a proposition is valid for the narrower field*?[72]

And again,

> Let us consider the following case: The narrower field G has the axioms A_G, and the totality of its purely logical consequences F_G; the broader field Γ, e.g. $A_G + A' = A_\Gamma$ or $A_\Gamma \supset A_G$, and thus the consequence ($F = Folge$)
>
> $$F_\Gamma = F_G + F_{A'} = F_{(G+A')}$$ [73]
>
> If some proposition or other [sc.: belonging to F_Γ] does not contain the compounds of the broadened operations, it is surely not obvious that it belongs to the F_G.

---

[68]*PdA* App. 439; *PoA* 419.

[69]Loc. cit.

[70]Loc. cit.

[71]'*Begriff*' in the original, but this clearly is a mistake.

[72]Loc. cit. (my emphasis).

[73]*PdA* App. 440; *PoA* 419–420 (our lettering follows the German original). Notice that the equivalence $F_\Gamma = F_G + F_{A'} = F_{(G\,+\,A')}$ (that is reported exactly like this also in Sch&Sch) contains an error: in general, it is not true that $F_G + F_{A'} = F_{(G\,+\,A')}$.

In summa, for Husserl the property of consistency of the expanded theory A_{Γ} does not imply the conservativity of A_{Γ} on A_G, while the theory of permanence tacitly and erroneously accepts this implication.

3.5 Passage Through the Imaginary

In the text *Der Durchgang durch das Imaginäre*, presented as the last part of the *Doppelvortrag* at Göttingen in the critical edition of 1971,[74] Husserl identifies a *property of axiom-systems* that ensures, in the situation mentioned above, the conservativity of the expansion: this is the property of 'definiteness'. On the basis of various definitions that Husserl gives,[75] we can identify this property with what is nowadays called *syntactic completeness* of a theory: "an axiom-system is relatively definite if every proposition meaningful according to it is decided under restriction to its field,"[76] that is if every formula (of the language of the theory) is either provable or refutable in it.[77] This identification has to be taken *modulo* the fact that Husserl and his contemporaries (in general, mathematicians and logicians up to at least 1917–1920) move in a *higher-order* logical environment, that is to say, they do not work under the now standard restriction to *first-order* languages and logic. Although we can find evidence in Husserl's writings of an *algorithmic* notion of *derivability* of a formula α from given axioms A according to certain pre-specified formal rules, one cannot seriously maintain that Husserl possessed the now standard and clear cut distinction between the *syntactic* notion of derivability from a (finite) set A of axioms and the *semantic* notion of truth in every structure in which the

---

[74]After much deliberation we have decided not to follow the reconstruction of the text as given in Sch&Sch where *Der Durchgang durch das Imaginäre* is preceded by a text titled '*Transcript from the Lecture* (*Abschrift aus dem Vortrag*)', in which Husserl takes up a crucial aspect of Hilbert's way of formulating axiomatic conditions for real numbers, known as *existential axiomatic*. The text *Abschrift aus dem Vortrag*, in Sch&Sch, concludes the first *Vortrag* while *Der Durchgang durch das Imaginäre* belongs to the second *Vortrag*, and is preceded by a long discussion of the concept of *mathematical manifold*. While we agree that this structure reflects the order in which Husserl must have presented his arguments, we consider presenting *Der Durchgang durch das Imaginäre*, which contains Husserl's peculiar solution for the problem posed by the theory of permanence, more convenient. Once this point has been clarified, we can look for a suitable mathematical counterpart for Husserl's concept of *mathematical manifold*. Finally, Husserl's use of existential axiomatics can be studied separately as one of the possible ways in which he thinks the system of axioms for a definite manifold can be established. (This is the topic of Appendix 6 below.)

[75]See also *Drei Studien zur Definitheit und Erweiterung eines Axiomensystems,* in: *PdA* App. 452–469, *PoA* 432–438 & 453–464.

[76]*PdA* App. 440; *PoA* 427.

[77]Cp. Tieszen 2005: "a 'definite' formal axiom-system appears to be a consistent and complete axiom-system, and a definite manifold is the system of formal objects, relations, and so on, to which a definite axiom-system refers" (4).

axioms A hold true.[78] Indeed, the autonomy of the syntactical moment of the theory with respect to the semantical one had not been made explicit at that time yet.[79] Husserl often uses in his definitions concepts like 'true', 'false' or 'logically entailed by' which suggest a semantical reading, whereas in the standard definition of syntactic completeness a strict notion of being syntactically derivable from an axiom-system is used. Moreover, Husserl also lacks a full awareness of the morphological aspect of the theory, and hence of the notion of formal language of a theory.[80] So, he refers to what is nowadays called a sentence in the language of the theory as "a proposition that makes sense with respect to the axiom-system ([*ein*] *für das Axiomensystem sinnhabender Satz*)"[81] or, he talks, with reference to the ontological counterpart of an axiom-system, of "a proposition falling within the domain".[82] As a consequence Husserl often oscillates between a characterization of definiteness that we would call *syntactical* and a characterization that we would call *semantical*: he seems to regard syntactical talk and semantical talk as coming more

---

[78]For an incisive discussion of these issues, as well as for a technically detailed and historically well documented study of the various notions of *completeness* which occurred in connection with the development of the axiomatic method in the late nineteenth and early twentieth century mathematics, see Awodey and Reck 2002. The authors take into consideration the origins and the progressive clarification and differentiation of this notion and of the related notion of 'categoricity', starting from the early 'tentative' characterizations to be found in the works of Dedekind and Peano as well as in Hilbert's *Grundlagen der Geometrie* (1899) and *Über den Zahlbegriff* (1900a), and then following its refinements by Huntington and Veblen, up to the clearer assessment in Fraenkel 1919 and in the sadly neglected Carnap 2000. It is a pity that Husserl's notions of 'definiteness' are not mentioned at all; though footnote 38 refers to Majer 1997 and Da Silva 2000 "for more historical and philosophical background, in particular involving Hilbert's relation to Husserl in this connection". – For the reasons given above in the text we cannot agree with the interpretation of *Ideas* I, §72 in Hartimo 2007, 298.

[79]As is well known, the clear distinction between syntactical and semantical aspects of a theory as well as the recognition of distinct logical levels (propositional, first-order, higher-order) as it is nowadays standard can be traced back to the years 1917–1919 that mark what is generally recognized as the "third phase" of Hilbert's foundational research. Hilbert's foundational program at that time is expressed in his Zürich talk *Axiomatisches Denken* (1917, published as Hilbert 1918). Hilbert appears to be strongly influenced by the *Principia Mathematica* of Russell and Whithehead published between 1910 and 1913. His foundational claim consists now in requiring (1) a strengthening of the logic of the *Principia* by way of formalization and axiomatization, (2) a reduction of the theory of cardinal numbers, of the reals and of set-theory to this strengthened logic and (3) a proof of consistency for this comprehensive great logic. By requiring (3) Hilbert distances himself from the viewpoint of logicism, for the proof of consistency had to be given in a new mathematical theory (the 'proof theory'). It is in this context that a metamathematical study of propositional and predicate logic begins. For example, soundness and completeness for propositional logic as a *separate* logical level are proven, and the concept and structure of a formal language are outlined in a rigorous way (cp. Abrusci 1978, 27–30).

[80]This does not contradict our claim (in Chapter 2, §§5–6) that Husserl's logical morphology foreshadows our concept of formal language, for we did not maintain that Husserl had the notion of formal language in the way that is today standard.

[81]Sch&Sch 111; *PoA* 438.

[82]*PdA* App. 441; *PoA* 428.

or less to the same thing. Consider the following three characterizations of the property of definiteness. The first appears to be more on the syntactical side:

> An axiom-system that delimits a field is said to be "definite" if every proposition intelligible on the basis of the axiom-system, understood as a proposition of the field, ... either ... follows from the axioms or contradicts them.[83]

The other two seem to be rather on the semantical side:

> Equivalent to this is the following statement: An axiom-system with a field is definite if it leaves no question related to the domain and meaningful in terms of this system of axioms open or undecided.[84]

> The field is definite if the truth and falsity of any such sentence is decided for the domain on the basis of the axioms.[85]

The question as to the interpretation of the property of definiteness is controversial and, as we shall see, different characterizations have been proposed for it. So let us try to provide further textual evidence for our interpretation of definiteness as *syntactic completeness* and add some clarifications. In the second *Vortrag* Husserl gives the following definition of a definite axiom-system:

> An irreducible axiom-system ... is definite ... when no independent axiom can be added which is constructed only from the concepts already defined (of course, also, none can be withdrawn, since otherwise the axiom-system would not be irreducible). [86]

By an 'irreducible axiom-system' Husserl means an independent one, that is a system which is such that none of its axioms follows from the remaining ones. As to the impossibility, on pain of inconsistency, of adding new axioms while preserving the independence of the system, this is a property which exactly corresponds – as is easily seen – to the property nowadays known as *maximality* or (sometimes) *saturatedness* of a formal system: informally speaking, a formal system T is maximal when it proves all that can be proved, on pain of inconsistency; that is, formally, when for each closed formula α of the language of the theory it holds that if α is not derivable from T then the system $T + \alpha$ is inconsistent (i.e., a contradiction is derivable from it). Now such a property of formal systems is well known to be equivalent in *classical* logic to syntactic completeness. It is worth noticing that Husserl seems to be fully aware of this equivalence. The passage quoted above continues as follows:

> *But I can also say*: I define an axiom-system, which formally defines a field of objects in such a way that every meaningful question for this field of objects finds its answer by means of the axiom-system; or that every proposition that is meaningful in virtue of the axioms ... either follows from the axioms or contradicts them.[87]

---

[83] *PdA* App. 457; *PoA* 438.

[84] Loc. cit.

[85] Sch&Sch 112; *PoA* 439.

[86] *PdA* App. 454; *PoA* 434; Sch&Sch 108.

[87] Loc. cit. (my emphasis).

Let us now return to Husserl's text, using the formal counterpart we chose to interpret the property of definiteness. If A_{Γ} is a consistent expansion of A_G and the proposition α, 'devoid of imaginary', can be proved from A_{Γ} then A_G does not prove $\neg\alpha$ and hence, because of the supposed 'definiteness' of A_G, A_G proves α. *Consistent expansions of definite axiom-systems are always conservative expansions.*[88]

Before considering the case more generally, Husserl gives an example. He considers a 'restricted arithmetic', *AR* (we can think of this as an *algebraic* theory[89] of the elementary operations on natural numbers) and a universal arithmetic *AU*. All formulae of *AR* – Husserl says – can be *reduced to equations* (even the non-identities can be aptly reduced to equations "just as when we understand $a < b$ as the equation b + u = a"[90]). But *AR decides all equations*: "each equation falling within that arithmetic is either valid on the basis of the axioms, or it is invalid on the basis of the axioms; that is, *either the proposition is a consequence of the axioms or it contradicts the axioms*."[91] Hence *AU* is *a conservative expansion of AR*:

> Accordingly we will state that for arithmetic the problem resolves itself in this way: Every proposition falling within the narrower, but deduced on the basis of the broader arithmetic, is an equation. Now every equation falling within the narrower arithmetic is either true (*richtig*) in it or contradictory (*widersprechend*) in it. An equation deduced within the broader field cannot be in contradiction with the axioms of the narrower field. Otherwise the entire broader field would be inconsistent. Therefore it is true.[92]

Generalizing, Husserl synthesizes the result of his reflections: a 'passage through the imaginary' is possible (1) if the imaginary can be *formally* defined in a consistent and comprehensive system of deduction, and (2) if the original field of deduction, when formalized, has the property that every proposition falling within that field is *decided* on the basis of the axioms of the field.[93]

The property of *definiteness* that we considered up to now is called by Husserl 'relative definiteness'. He also considers another kind of definiteness which he calls '*absolute' definiteness* or sometimes 'improper (*unechte*) *completeness* in a Hilbertian sense'. ("Absolutely definite = complete, in *Hilbert's* sense.")[94]

In essence, an axiom-system is *absolutely definite* when it contains an *axiom of closure* (*Schliessungsaxiom*) analogous to the *axiom of completeness* that Hilbert includes in his (categorical) axiomatic characterization of the system of real numbers: "it is not possible to add to the system of numbers any collection of

---

[88]Ortiz Hill 2002, 89–94 discusses at length this issue though she does not expressly interpret the property sought by Husserl as *conservativity* of a theory, nor does she provide a formal equivalent for the notion of definiteness of a theory.

[89]On the essentially *algebraic-equational* nature of the 'arithmetics' under consideration, see below.

[90]*PdA* App. 440; *PoA* 428. But notice that strictly speaking b+u=a is no longer an equation since the variable u is (tacitly) *existentially* quantified.

[91]*PdA* App. 441; *PoA* 428. (my emphasis).

[92]Loc. cit.

[93]Loc. cit.

[94]*PdA* App. 440; *PoA* 427.

things such that, in the reunited collection, the previous axioms are satisfied; that is . . . the numbers form a system of objects that cannot be expanded in such a way that the previous axioms remain valid."[95] In Husserl's general formulation, "by such and such axioms the field is determined, and no other axioms are valid for it."[96] For him this is a *negative* axiom to which no interesting property of the axiom-systems corresponds at all, because every system can be made complete by adding such an axiom. "Such 'completeness' is, of course, not something peculiarly characteristic [in the intrinsic sense] of axiom-systems."[97]

In the (studies for) the second *Vortrag* Husserl writes:

> Finally, I further distinguish *relatively* and *absolutely* definite axioms-systems. An axiom-system is *relatively definite* if, for its domain of existence it admits no additional axioms, but it does admit that for a broader domain the same, and then of course also new, axioms are valid. New axioms, since the old axioms alone in fact determine only the old domain.[98]

Here[99] Husserl also rephrases this distinction as that between "extra-essentially complete (*ausserwesentlich vollständige*)" and "essentially complete (*wesentlich vollständige*)" axiom-systems. He explicitly claims to have considered in his investigations only the first kind of axiom-systems, whereas the concept of absolutely complete axioms system had remained outside of his consideration. He writes:

> From these considerations we easily arrive at axiom-systems that are "complete" in Mr. Hilbert's sense. The axiom-systems considered up to now which I called "definite", I shall henceforth call "extra-essentially complete," in contrast to those that are complete in Hilbert's sense, which I shall call "essentially complete." This latter concept remained hidden to me, since for my purposes everything was accomplished by means of extra-essential completeness.[100]

In the "Passage through the Imaginary" Husserl observes that absolute definiteness implies, in an obvious way, relative definiteness: "it is certainly true that such an axiom-system, closed in an exterior and spurious manner, already has the property which we had in mind: namely, it can be read off[101] of each proposition whether it is or is not a consequence of the axiom-system."[102] He then asks whether there are axiom-systems that do not contain an axiom of closure and that

---

[95]Hilbert 1900a, 183. Cp. Webb 1980, 84.

[96]*PdA* App. 442; *PoA* 429.

[97]Loc. cit.

[98]Sch&Sch 102; *PoA* 426.

[99]*PdA* App. 450–457; *PoA* 432–438; Sch&Sch 107–111.

[100]*PdA* App. 455; *PoA* 436; Sch&Sch 110.

[101]This 'can be read off' adds a connotation of *effectiveness*, that in itself is not part of the formulation of the property of syntactical completeness. On the other hand, it is true that if an elementary theory *T* is recursively axiomatized (i.e. has a [semi]decidable set of specific axioms) and syntactically complete, then *T* is *decidable* (i.e. there is an effective method that allows to establish, for any given closed formula of its language, whether or not it is provable in *T*).

[102]*PdA* App. 442; *PoA* 429–430.

nevertheless, *on the basis of their particular nature*, are capable of deciding every formula, in other words, whether there are axiom-systems that are *relatively* but *not absolutely* definite. His answer, already partially anticipated in the concrete example of 'definite' system, is affirmative: not only the arithmetics of the natural numbers, but *all arithmetics* (of the integers, etc.; also that of the reals) are examples of such systems. The reasons Husserl advances for this statement are, essentially, three.

(i) The question of the formal decidability of generic 'algebraic' formulae can be reduced to that of the formal decidability of *equations*.
(ii) An *equation* has the form '$t = s$', where 't' and 's' are 'operative constructs' (i.e. *terms*) that can be formally 'calculated' on the basis of the axioms of the theory, which constitute the 'rules of calculation'. Comparing the 'normal forms' of 't' and of 's' allows us to decide the equation formally: if they are identical, the equation follows from the axioms, if not, the equation is in contradiction with them and hence its negation follows from the axioms.

> For a numerical equation to obtain (*bestehen*) means, of course, that given the execution of operations in the sense of the axioms the identity $a = a$ is produced. Every numerical equation is true if it can be transformed into an identity, and otherwise false. Every algebraic formula is, then, also decided, for it is decided for each numerical case.[103]

Let us digress here for a moment. Speaking of syntactically complete ('definite') arithmetical theories after Gödel's limitative results might sound puzzling. Actually, to try to make sense (as far as possible) of what Husserl is saying, one has to keep in mind that – as is evident from the above quotation – *in this context* by 'theory' of a certain numerical domain he means something like an *algebraic* kind of theory, a 'theory of operations', in which statements are (or can be reduced to) equalities and possibly inequalities between terms. In this sense of 'theory' it is in fact possible to design interesting syntactically complete arithmetical theories – at least in some specific cases, if not in general. It is worth briefly mentioning here what is perhaps the most famous example, the first-order theory RCF of *real closed fields* (also called *elementary algebra*), whose privileged model (but by no means the only one) is the field of the real numbers. The language of RCF contains the identity predicate $=$, a binary predicate $\leq$ for the order relation, the binary function letters $+$ and $\cdot$ (for addition and multiplication) and the individual constants 0 and 1. Its specific axioms feature the usual finitely many axioms for ordered fields, plus one axiom saying that every positive element has a square root, and (infinitely many) axioms saying that every polynomial of odd degree has a root. Now, as proved by A. Tarski,[104] RCF admits elimination of quantifiers, that is: to every (open) formula A in the language of RCF a quantifier-free formula A' can be effectively associated, such that A and A' are equivalent in RCF (thus, basically,

[103] *PdA* App. 443; *PoA* 430.

[104] The relevant part of the work had already been completed by Tarski around 1930; yet the result appeared in print only many years later, first as a technical report (1948) and finally in Tarski 1951.

Husserl's above claim (i) is met). As a consequence, the theory RCF is syntactically complete and, in turn, decidable. Any sentence in the first-order language of RCF is true in an arbitrary model of RCF if and only if it is true in the reals.[105]

Let us now come back to Husserl's arguments in support of his claim that all arithmetics are examples of relative or extra-essentially complete axiom-systems.

(iii) If we consider the matter from the point of view of the structures underlying the theories, the justification of the fact that the arithmetics are definite theories is articulated as follows: to each arithmetic corresponds a well-determined structure of the field of objects (manifold) axiomatized by it; that is to each arithmetic corresponds a different numerical system, and numerical systems are considered as inductively generated manifolds,[106] that is, structures such that (1) a number of initial elements is specified, which belong to the domain without any further condition, (2) certain procedures are specified to generate new elements from given ones, (3) the domain so identified is the smallest among all those that satisfy the two first conditions.

> Every arithmetic, regardless of how it is restricted – whether it has reference to the whole positive numbers, or to the whole real numbers, or to the positive rational numbers, or to the rational numbers in general, etc. – every arithmetic is defined by an axiom-system such that, on its basis, we can prove: every proposition in general that is constructed exclusively of concepts which are established as valid by the axioms (or are axiomatically admitted), every such proposition falls in the field, i.e. it is either a consequence of the axioms or contradicts them. The proof of this assertion lies in the fact that every defined operational formation is a natural number and that each natural number stands to every natural number in a relation of order determinable on the basis of the axioms.[107]

When a formal system is such that "it can be shown on the basis of the axioms that every object of the field reduces to the group of the numerical objects," then – Husserl states – such a system is *definite*. And he adds: "whenever, for example, each defined proposition is reducible to an equation or to the >/< between numerical objects, the axiom-system is definite."[108] So definiteness is for Husserl an internal property (*innere Eigenschaft*) of an axiom-system, intrinsically connected with the kind of structure determined by the axiom-system. If the domain of the latter is inductively generated starting from a finite number of initial objects by means of pre-specified generation procedures, an axiom-system for such a structure is definite.

---

[105]It may be helpful to observe that, as far as the 'ontological counterpart' of RCF is considered, there are some drawbacks which Husserl would not have liked, as we shall see in Section 7 below. Indeed, while it is obviously true that the privileged model of RCF (the ordered field of reals) contains the standard structure of natural numbers as an 'embedded image' (that is, as a substructure), it is known that such an image cannot be defined within RCF: the set of standard natural numbers (and the successor function as well) cannot be formally defined within RCF.

[106]The construction of the number system in the *Philosophy of Arithmetic* can serve as a model. Cp. Chapter 1 §11 ff.

[107]*PdA* App. 442–443; *PoA* 430.

[108]*PdA* App. 443; *PoA* 431.

Still more explicitly, this position is restated in Husserl's notes on a paper presented by Hilbert on November 5, 1901.[109] As Majer correctly observes,[110] the notes begin as a reproduction from memory of Hilbert's paper, but very soon Husserl breaks off and reports an objection Hilbert made to him during the discussion following his talk. This objection has not been noticed, Majer remarks, because, being among the notes taken of Hilbert's paper, it had been considered as part of them.

Husserl writes: "*Hilbert's* objection: — Am I justified in saying that every proposition containing only the whole positive numbers is true or false on the basis of the axioms for whole positive numbers?"[111] On our interpretation this means: "Am I justified in saying that the theory of natural numbers, and therefore all arithmetics, are *definite*, that is *syntactically complete*?" This question is taken up again, and answered, on the next page:

> How do I know that? Every direct operational combination (*Operationsverbindung*), however often it may contain each operation, is *equal to a number*... Therefore every proposition which asserts two algebraically general, closed expressions to be equal – and likewise every mixed equation built up from algebraic and number signs – will of course have to be necessarily true or false on the basis of the axioms. For: whichever group of numbers I may substitute for the a, b, c,... p in a formula, there is always one determinate number for each side of the equation. And, indeed, on the basis of the axioms. If it is satisfied for all possible combinations of numbers, then the formula is valid. If not, it is not ... It suffices that I can demonstrate from the axioms that every expression is a number, and consequently it is self-evident that two expressions either always represent the same number or different numbers.[112]

We take this to mean that the arithmetics are syntactically complete because every expression α can be reduced to an equation and the axioms of the theory "calculate" all equations.

In the studies for the second *Vortrag* Husserl explicitly ascribes *relative* definiteness to all arithmetics – with the exception of the arithmetic of the reals, for the latter system is intended to be categorical. This claim might seem to clash with what we said above, but in fact there is no conflict. Husserl is not thinking here of an *algebraic* theory of the basic operations on the reals; he is thinking of a full fledged theory – like the one put forward by Hilbert – capable of characterizing up to isomorphism the continuous, uncountable structure of the reals.

> *Relatively definite is the sphere of the whole and the fractional numbers, of the rational numbers, likewise of the discrete sequence of ordered pairs of numbers* (*complex numbers*). I call a manifold *absolutely definite* if there is no other manifold which has the same axioms

[109] "*Notes on a lecture by Hilbert* (*Notizen über einen Vortrag von Hilbert*)" (*PdA* App. 444–447; *PoA* 464–468).

[110] Majer 1997, 39. For a contrasting interpretation of the very same point see Sch&Sch 89.

[111] *PdA* 445; *PoA* 465. Cp. Webb 1980, 84–85.

[112] *PdA* App. 446; *PoA* 466–467.

> (all together) as it has. Continuous number sequence, continuous sequence of ordered pairs of numbers.[113]

> The essentially complete axiom-systems . . . form . . . the outermost sphere within which the expansion of extra-essential axioms can move, by leaving the original axioms system unaltered.[114]

These two quotations provide further evidence for our thesis that, on the one hand, Husserl's approach to the number concept is of a "genetic" kind, for he thinks to obtain the consistent multiplicities of the various number systems by means of successive expansions starting from the arithmetic of natural numbers[115] and that, on the other hand, Husserl's *Doppelvortrag* represents – as it were – his contribution to Hilbert's methodological issue at that time, that is the axiomatic foundation of the system of the reals. Husserl does not try to prove directly the consistency of arithmetic of the reals but to prove that the stepwise expansion of the consistent arithmetic of natural numbers remains in each step conservative upon the old domain. The set of calculation rules for natural numbers can be expanded so as to interpret a broader domain, for example that of the wholes: one joins new elements and axioms for the new elements. We obtain a conservative expansion, conservativity being implied by definiteness, a property that, according to Husserl, pertains to each arithmetic. Husserl's idea is that the theory we begin with should be expanded until it contains all numerical systems. Both universal arithmetic and each of the specific arithmetics, with the exception of the full arithmetic of the reals, are extra-essentially complete axioms systems, they leave open the possibility to be further specified. For instance, the domain of the arithmetic of the wholes is completely determined by the rules for the operation for these numbers. To the domain delimited by the axioms for the wholes no new axiom can be joined.

> A "definite" axiom-system leaves for its operational substrate absolutely nothing open with respect to the operations defined. If it were to leave anything open, there would in fact be relations which are not true or false on the basis of the axioms. *And yet something remains open. Namely, a restricted arithmetic. And a domain of deduction is restricted in an analogous sense, if not all operational constructs* that remain free on the basis of the general laws of operations are defined, and then of course the specific laws of operation [relative to these constructs] are not introduced either.[116]

Once determined the operational forms that have been left open by a particular specific arithmetic only two cases are possible:

(i) "Either we have set up a series of axioms [*Festsetzungen*] in such a way that possible operational constructs are still left open and yet, so far as we have

[113]Sch&Sch 102; *PoA* 436–437.

[114]*PdA* App. 455; *PoA* 435; Sch&Sch 109.

[115]Cp. Cavaillès 1938: "Ainsi procédait-on dans l'école de Weierstass, ainsi Kronecker reconstituait-il toute l'analyse à partir du nombre entier 'seul créé par le bon Dieu'" (84).

[116]*PdA* App. 456; *PoA* 436–437.

defined it, the axiom-system is definite."[117] Intuitively, it is possible to add further formal axioms, but not with respect to the already determined operations

(ii) "Or else there remain open no further operational results (*Operationsergebnisse*) whatsoever – *nota bene*, none that are possible compatibly with the general basic laws and the specific laws already defined",[118] that is it is not possible to expand the formal system any further.

In the first case the axiom-system is extra-essentially complete, in the second case it is essentially complete.

3.6 On Different Interpretations of Husserl's Notion of Definiteness

A number of different interpretations of Husserl's notion of *Definitheit* has been proposed in the literature. Let us consider some of the most significant ones and outline our motives not to share them. This will throw some further light on Husserl's *desiderata*, or so we hope.

3.6.1 Husserl's Two Notions of Definiteness

In his 2000 da Silva interprets 'absolute definiteness' of a theory as syntactic completeness *tout court*, and 'relative definiteness' of a theory as syntactic completeness *restricted to a specified set of formulae*, where a theory **T** is 'relatively definite' with respect to a set Σ of formulae if and only if for every $\alpha \in \Sigma$ it is the case that either **T** proves α or **T** proves $\neg\alpha$. The author's distinction is intended to apply to a given formal theory **T**, over a certain language L, which is not categorical (hence does not have only one model, up to isomorphism) but which possesses a *privileged* model **D** (called the 'formal domain' of **T**). Then, such a **D** determines a sublanguage $L_{\mathbf{D}}$ of L, which is obtained by restricting quantification to the objects of the formal domain **D**, and this $L_{\mathbf{D}}$ determines in turn the *specified set of formulae* to which syntactic completeness has to be restricted. A serious objection to this interpretation of the notion of relative definiteness (which the author himself seems to consider, although he decides not to discuss it, for the sake of simplicity of exposition) concerns the question how $L_{\mathbf{D}}$ can be precisely determined *as a sublanguage of* L. One might think of a monadic formula $\delta(x)$ of L defining the objects of **D**, so that $L_{\mathbf{D}}$-formulae would in this case be just those formulae of L in which quantification is restricted to $\delta(x)$ (i.e. all quantifier occurrences are of the

[117] *PdA* App. 454; *PoA* 436–437.

[118] Loc. cit.

form: '$\forall x(\delta(x) \rightarrow \cdots)$ and '$\exists x(\delta(x) \wedge \cdots)$'. The problem is that such a L-formula δ(x), in most cases, does not exist (for instance, if **T** is first-order Peano arithmetic PA and **D** is the standard model of PA, it is known that no L(PA)-formula δ(x) exists which defines the set N of standard natural numbers in every model of PA). Da Silva also criticizes the interpretation of 'relative definiteness' as syntactic completeness *tout court*, arguing as follows: suppose that **T** is syntactically complete, and that **T** does not prove "there are imaginaries." Then, by syntactic completeness, **T** must prove that "there are no imaginaries," and so no extension **T'** of **T** can exist which proves "there are imaginaries," thus trivializing the whole problem under discussion. But, again, we can object to this argument as above, by observing that nothing ensures us that the sentence "there are imaginaries" is expressible in the language of **T**.

3.6.2 *Husserl's* **Definitheit** *and Hilbert's* **Vollständigkeit**

In *Ideas* §72 Husserl mentions the "close relationship of the concept of definiteness to the 'axiom of completeness' introduced by Hilbert for the foundation of arithmetic."[119] In *Formal and Transcendental Logic* he says "even if the innermost motives that guided him [Hilbert] mathematically, were inexplicit, they tended essentially in the same direction as those that determined the concept of the definite manifold."[120] In her contribution to the topic we are discussing[121] C. Ortiz Hill quotes these remarks and goes on to identify these two concepts *tout court*. She focuses rather on the problem of the epistemological and philosophical positions that are assumed by the two authors. She writes in a note: "First of all, for complete and completeness Husserl uses the German words "*definit*" and "*Definitheit*" in the place of Hilbert's '*vollständig*" and "*Vollständigkeit*'. Since in the passages cited above Husserl maintains that his concept of *Definitheit* is exactly the same of Hilbert's *Vollständigkeit*, I have tried to avoid the terminological confusion by translating Husserl's terms with the more familiar 'complete' and 'completeness,' although Husserl translators have understandably chosen 'definite' and 'definiteness'."[122]

This terminological assimilation hides the significant conceptual differences between Husserl's concept of "definiteness of a theory" and Hilbert's concept of "completeness," introduced in *Über den Zahlbegriff* as "completeness of the system of axioms for real numbers." The system of axioms for real numbers is categorical; all its models are mutually isomorphic. In the discussion above we have tried to give textual evidence for the claim that, according to Husserl, the property of categoricity belongs rather to those formal systems that are absolutely definite: Husserl's "relative definiteness" (the syntactical completeness of a theory T, on our

---

[119] *Ideas* 164, 17.

[120] *FTL* 85.

[121] Ortiz Hill 1995. In: Hintikka 1995.

[122] Ortiz Hill 1995. In: Hintikka 1995, 161, note 2.

interpretation) *is not meant to imply* (and in fact it does not imply) that T is also "categorical," while it can be demonstrated that the models of a *syntactically complete* theory are all structurally very similar, and the problem arises as to how to interpret this structural similarity.[123]

Moreover, it is questionable to maintain that in the *Doppelvortrag* "Husserl searched for answers regarding the consistency of arithmetic".[124] Hilbert proposed the requirement of a proof of consistency for arithmetic in what is known as the "first phase" of his foundational investigations (1898–1901), explicitly in his address to the second International Congress of Mathematicians in Paris in 1900, titled *Mathematische Probleme*. "But above all I wish to designate the following as the most important among the numerous questions which can be asked with regard to the axioms: *To prove that they are not contradictory, that is, that a definite number of logical steps based upon them can never lead to contradictory results*."[125] However, we have tried in our discussion above to provide evidence for the thesis that Husserl is inclined to take the consistency of the axioms of arithmetic for granted. The thesis that Husserl proposes in the *Doppelvortrag* is a conditional claim: "if **T** is consistent and syntactically complete (definite) then every consistent extension of **T** is conservative, so that the transition through the imaginary is justified."

Regarding the proximity of Husserl's notion of "passage through the imaginary" to Hilbert's "ideal elements," it must be said that Hilbert starts talking about "ideal elements" in what is known as the "fourth phase" of the foundational research (1920–1924) in which "Hilbert's Program" takes on its proper shape. In a paper presented in Leipzig in 1922 entitled *Die logischen Grundlagen der Mathematik* (Hilbert 1923), Hilbert points to finitary mathematics as that part of mathematics that "has a concrete content", that is that operates concretely with symbols and does not use the procedures and principles criticized by predicativists and intuitionists. It is in this context that he starts talking about "finitary logic" (the logic of finitary procedures) and of "ideal elements." In 1925 in Münster Hilbert presents his paper *Über das Unendliche* (Hilbert 1926), where he explicitly speaks of ideal elements. Here finitary mathematics is said to be that part of mathematics that can be rightfully considered as "secure." It does not need a justification, but must itself serve as justification for infinitary mathematics, which is that part of mathematics that deals with actual infinity, lacks a concrete content and is moreover a possible source of contradictions. Infinitary instruments are acknowledged as "useful": they are used to prove real expressions. However, they must be justified, that is we must demonstrate that their use does not lead to a contradiction. In the same way, we must demonstrate that *ideal mathematics is conservative with respect to real mathematics*. – Now, without wanting to detract anything from the hypothesis

[123]See below §§6–8. For the reasons given above in the text we cannot agree with Hartimo's understanding of the property of *relative definiteness* of an axiom-system in the *Doppelvorstrag* as its categoricity, cp. Hartimo 2007.

[124]Ortiz Hill 2005. In: Hintikka 1995, 145.

[125]Hilbert 1900b.

that "Hilbert's deep underlying reasons for formulating his axiom of completeness were basically the same as those which had led Husserl to formulate his own concept of completeness",[126] given the difference in dating (the *Doppelvortrag* was written 25 years before Hilbert started talking about "ideal elements"), we think it more likely that the source of Husserl's idea is rather the methodological hypothesis (which surfaced repeatedly during mathematical investigations in the nineteenth century) that analysis should be conservative with respect to number theory, that is that the numerical expressions that are provable in analysis would in any case already be provable without using the transfinite.

3.6.3 *Did the* Doppelvortrag *Ever Confront the Problem of Semantic Completeness?*

In his 1997 Majer maintains that in the *Doppelvortrag* Husserl discusses the following question: "Under what conditions does the *truth* of a theory follow from its *consistency*?" The problem is relevant and hence worth to be considered. Majer's contention can be recast as follows: Husserl was puzzled about Hilbert's idea that the syntactic consistency of an axiom-system is sufficient to guarantee the existence of a system of entities formally characterized by those axioms. It seems that Majer first endorses the following rather strong claim: it can be conjectured that Husserl foresees the necessity to *prove* – as one would say in present-day terminology – the correspondence between the consistency of a theory and the existence of a model for that theory; in short, the necessity to prove a theorem of semantic completeness, at least for theories that axiomatize numerical systems. Majer writes: "If it could be proved that the truth of a theory follows from its consistency the search for an indubitable foundation of mathematics would be a tremendous step forward . . . But, as you know, these things are not so easy. And Husserl was quite aware of this. . ."[127] Majer then endorses a weakening of the above claim, namely, that Husserl *criticizes Hilbert for omitting to make the notion of truth relative to 'truth in a structure'*. Majer seems to find evidence for this claim in the observations Husserl had added to an exchange of letters between Hilbert and Frege[128] about Hilbert's *Grundlagen der Geometrie*. Of special interest for us are Frege's letter from December 27, 1899 and Hilbert's answer from December 29. Frege says:

> Axioms I call propositions that are true but are not proved. . . From the truth of the axioms it follows that they do not contradict each other; so this does not require any further proof.

[126]Ortiz Hill c, Husserl and Hilbert on completeness. In: Hintikka J 1995, 144.

[127]Majer 1997, 38.

[128]This exchange is contained in Frege 1976, 55–80; Husserl's annotations are published *PdA* App. 476–451; *PoA* 464–673. For a critical assessment of the Frege-Hilbert debate on this point see Blanchette 1996.

> Definitions, too, must not contradict each other. If they do they are mistaken. The principles of defining have to be such that if we follow them no contradiction can arise.[129]

Husserl comments:

> Frege does not understand the sense of Hilbert's 'axiomatic' foundation of geometry. Namely, that this is a purely formal system of conventions which coincides, as to the form of the theory, with the Euclidean.

He then presents excerpts from Hilbert's answer[130] and writes:

> The proposition "from the truth of the axioms follows that they do not contradict each other" interested Hilbert a lot, because when he thinks, writes and talks about such things he says exactly the opposite: "If the axioms that are arbitrarily fixed do not contradict each other ... then they are true, then the things defined by the axioms exist. For me this is the criterion of truth and of existence..." In this sense we speak of "the existence of real numbers," of the "non-existence" of the system of all cardinals (*Mächtigkeiten*).

Majer seems to read Hilbert and Husserl in the following way: while Hilbert took the consistency of the axioms to be a sufficient condition for there being a structure that verifies those conditions, Husserl, on the contrary, thought that if we formalize a concrete theory and then expand it formally in a consistent way, then one should *prove* that to the expanded consistent theory there still corresponds a system of entities, for this does not immediately follow from the consistency of the axioms. However, if some further constraints could be imposed on the theory, for example definiteness – and all arithmetics are for internal reasons definite – then one can take semantic completeness for granted as Hilbert does.

As opposed to this interpretation we tried to provide evidence for the following thesis: in the *Doppelvortrag* Husserl reaches a position that is very close to Hilbert's. He also assumes that the mathematical existence of a system of entities is *guaranteed* by the consistency of the theory that implicitly defines that system. More generally, Husserl at this point does not have a clear vision of semantic completeness as a problem, but rather tends to take it for granted *tout court*. In the (preparatory studies for) the second *Vortrag* Husserl makes a remark that he does not develop into an argument: "Field of the axioms system. We restrict ourselves to axiom-systems that have a field. (Why [don't we say] directly: collection of objects that satisfy the axioms [?])."[131] He might as well have said: "we restrict ourselves to *consistent* axiom-systems", thereby identifying the consistency of an axiom-system *tout court* with its having a model. He goes on to say: "All remains correct it we simply take 'field' in the natural sense of the term: objects which satisfy the axioms."[132] So he simply takes the correspondence between consistent theory and structure for granted.[133]

---

[129]Frege 1976, 63.

[130]Frege 1976, 66.

[131]*PdA* App. 457; *PoA* 437–438; Sch&Sch 111.

[132]*PdA* App. 457; *PoA* 438 n.; Sch&Sch 111 n.

[133]I do not even agree with Majer's attempt to interpret Husserl's answer to Hilbert's objection, and, more generally, Husserl's conception of arithmetics as essentially algebraic or equational

To sum up: the problem that Husserl addresses in the *Doppelvortrag* is exactly what he himself announces: "how can we justify the use of the imaginary in calculations?" He reformulates the question as follows: under which conditions can we expand a given theory with imaginary concepts and still be certain that contradictions will not be implied by this expansion? And his answer is: if the given theory is syntactically complete and the expansion is conservative with respect to the old domain then the use of the imaginary in calculations cannot yield contradictions.

3.7 More on the Conservativity of Expansions

In the preparatory studies for the second *Vortrag*[134] Husserl takes into account some further aspects of the question concerning the possibility to expand an axiom-system in a *conservative* way.

Let A be a given axiom-system, and A_w a formal expansion of it (*erweitertes Axiomensystem*). In symbols:

$$A \subseteq A_w.$$

An expansion should satisfy, according to Husserl, the following conditions:[135] Given that $A_w = A + B$, the new axioms

1. have to be consistent:

   $not(B \vdash \alpha \wedge \neg\alpha)$;

2. have to formulate assertions that are left open by the old axioms:

   for all $\beta \in B, not(A \vdash \beta)$ and $not(A \vdash \neg\beta)$.

theories, as "an axiomatic description" of the intuitively given numerical field. As a matter of fact the latter is introduced in these terms only in *PdA*. Majer writes: "if we restrict the axiomatization of numerical relations to numerical equality and inequality and to the two operations of [*Verbindung*] and [*Teilung*] of whole positives, a deductively complete axiomatization of the whole positives becomes possible". However, we must heed the fact that Husserl in the *Doppelvortrag* does not mention at all the intuitive operations of *Verbindung* and *Teilung*. The latter are introduced in *PdA* in a completely different context. An ideal continuity of the *Doppelvortrag* with *PdA* consists rather in the following fact: *PdA* provides the "rules of calculation" for the theory of finite cardinals and the *Doppelvortrag* investigates under which conditions the expansion of that theory remains conservative. For these reasons I do not share Majer interpretation of the *Doppelvortrag* as a solution to a conceptual leap in *PdA* between the intuitive origin of the numerical field and its properly computational side. Though this leap does indeed exist in the *PdA*, I fail to see how this can be legitimately said to be the problem that Husserl explicitly addresses in the *Doppelvortrag*.

[134] *PdA* App. 452 ff; *PoA* 433 ff. n.; Sch&Sch 107 ff.

[135] *PdA* App. 453; *PoA* 434 n.; Sch&Sch 108.

Moreover,

3. A_w, that is $A + B$, has to be consistent (note that 3 trivially implies 1).

He then discusses the question of conservativity again, thereby bringing to light some further aspects: *if we derive from* A_w *an assertion that refers purely to the objects* A, does it hold that *this assertion was already derivable from* A?[136]

As we saw, A_w is a conservative expansion of A iff A_w is an expansion of A and, for each assertion α of the language of A, it holds that if $A_w \vdash \alpha$ then it already holds $A \vdash \alpha$. Husserl would say: "A_w is a conservative expansion of A iff A_w is an expansion of A and, for each assertion α that 'has a sense' for the field determined by the 'narrower' axiom-system A, if α follows from the 'expanded' system A_w then α already follows from the 'narrower' system A."

Husserl's positive answer to the question italicized above is articulated as a distinction of cases which we can reconstruct as follows:

(1) Either one already knows that A_w is a conservative expansion of A, and hence α was already derivable from A ("[either] I know quite in general that every assertion which has a sense for the narrower domain . . . must be a consequence of the A . . ."[137])
(2) Or else one already knows that A is *definite* (syntactically complete). But then, from $A \subseteq A_w$ together with the consistency of A_w, the assumption that $A_w \vdash \alpha$ yields that *not*($A \vdash \neg\alpha$), and hence, *in virtue of the property of syntactic completeness of* A, the conclusion that $A \vdash \alpha$ follows also in this case ("or I know apriori . . . that this assertion belongs to the class of assertions that must be decided a priori by means of the A . . ." [138])

Generally, use of a broader axiom-system in order to derive propositions of a narrower one is allowed if we have a property at our disposal that guarantees that each assertion that "has a sense" in the narrower domain is also *decided* there, that is, it follows from the axioms or is in contradiction with them (*definiteness*).[139]

It is interesting to note that Husserl also considers the two notions of *definiteness* and *conservativity relativized to classes of assertions*. *De facto*, to obtain conservativity of an expansion A_w of A with respect to a *restricted class* Γ *of assertions*, it is sufficient to assume that A is Γ−definite, that is, syntactically complete relative to assertions which belong to the class Γ.

> The inference from the imaginary is permitted in a singular case or for a class, if we can know in advance and can see that for this case or for this class the inference is decided by the narrower system.[140]

[136] *PdA* App. 453; *PoA* 433; Sch&Sch 108.
[137] Loc.cit.
[138] *PdA* App. 453; *PoA* 433–434; Sch&Sch 108.
[139] *PdA* App. 456–457; *PoA* 437; Sch&Sch 111.
[140] *PdA* App. 457; *PoA* 437; Sch&Sch 111.

3.8 Definite Manifolds

The *Husserliana* edition of the *Philosophie der Arithmetik* contains two further studies on the problem of definiteness of manifolds.[141] The text is worth analyzing, for it deals, though in a rather fragmentary and tentative style, with a number of extremely interesting problems and conceptualizations as regards definiteness and conservative expansions with respect to manifolds. The question under scrutiny in these studies is this: which conditions have to be satisfied by a manifold if it is to be definite? In which sense is it possible to expand a definite manifold?

Husserl considers here the possibility of effecting a conservative expansion of a definite manifold. He asks, in particular, under which conditions assertions (and classes of assertions) relatively to the expanded manifold "can be transferred" to the narrower one and *vice versa*.

If the expansion of a manifold is to be conservative, the 'narrower' manifold (the one to be expanded) should be – to put it informally, paraphrasing Husserl – *embedded as image* (*Gebilde*) in the 'expanded' one, and at the same time it must be *left unchanged* by this embedding. Husserl tries to make this idea precise as follows.

Let M_0 be a certain definite manifold and M_E a conservative extension of it. M_E is made up of M_0 plus certain new elements, in symbols:

$$M_E = M_0 + e_n.$$

"The expansion to M_E must not disturb M_0 in what it is by itself, and above all, it must not specialize it, that is the axioms (*definitorische Bestimmungen*) for M_E must be a *mere* expansion of those for M_0."[142] Thus M is an expansion of M_0 if M_0 does not undergo, in M, any "specialization", that is, if it does not receive any determination that results in new propositions both for M and for its elements and constructs.[143]

Now let L_0 (resp. L_E) be the laws (*Gesetze*) that are valid for M_0 (resp. M_E), L_n be the laws that determine relations between the 'new' elements e_n, and L_{0n} the laws that determine relations between the old and the new elements. Then – Husserl says – it must hold that

$$L_E = L_0 + L_n + L_{0n}$$

This means that "the assumption that the manifold [M_E] reduces to the partial manifold M_0 ... should have the consequence (*bedingen*) that the L_E reduce to the L_0 without any further determination for M_0 thereby resulting."[144]

---

[141] *PdA* App. 458–469; *PoA* 453–464.

[142] *PdA* App. 459, *PoA* 454.

[143] *PdA* App. 461, *PoA* 456.

[144] *PdA* App. 460, *PoA* 455.

The idea is that the laws that turn out to be valid for the manifold M_0 *as an image in* M_E should be the same as the laws valid for M_0 'when it was not an image'. Husserl gives here also several examples of how certain forms of assertions that are valid for the expanded manifold "move" into the narrower one: if one, for example, proves relatively to M_0 as image that some A are B, then this surely also holds for the 'independent' M_0. Analogously, if one proves that there are some A.[145]

The simplest case of a conservative expansion, Husserl observes, turns out to be this: "formally the same laws hold both in the expanded domain and in the narrower one. L_0, L_n , L_{0n} *then have the same form.* By limitation [from the domain of the manifold M_E] to the domain 0 [sc.: M_0], L_0 proceeds immediately from L_E and nothing else does."[146] In this particular case, one just acquires more generality by the expansion, and immerging one structure in another simply consists in joining new objects. For the latter the same laws hold, or more exactly, laws of the same form hold as those true of the narrower domain.

The most interesting case – it is the intended exemplification of the whole discussion – is the case of numerical systems. As we saw, Husserl's idea is to progressively extend calculation rules for the *equational* theory of finite cardinal numbers until the expanded theory finally embraces the theories of all numerical systems (integers, rationals, and reals too). From the side of the ontological counterpart of the theories, the narrower structure (natural numbers) turns out to be embedded as an image in the immediately broader one.

> The series of the positive wholes number is a part of the series of numbers that is infinite at both ends. This in turn is part of the two-fold manifold of the complex numbers. The system of the positive whole numbers is defined by certain elementary relations. In this latter nothing is modified through the expansion of the number series. No new elementary relations are added, but rather only new elements and relations between the new and the old. The laws of the expanded domain include those of the narrower one, but in such a way that for the old domain no new laws are established.[147]

The above issues, like those dealt with in the previous sections, touch very interesting yet delicate points from the metalogical point of view, and they easily call to mind certain notions and results familiar from contemporary logic. Yet, here as elsewhere, some caution is necessary, just to avoid forcing or misinterpreting Husserl's views. Let us consider the above concept of "conservative expansion", whereby a narrower manifold M_0 is 'embedded as image' into an extended one M_E. It is tempting to ask which 'technical' notion from model theory provides us with an (at least reasonably approximate) interpretation of that concept. A first candidate is, of course, the basic notion of *extension.* Given two structures M_0 and M_E of the same type or signature (that is, two structures for the same elementary language), M_E is called an extension of M_0 (equivalently: M_0 is called a *substructure* of M_E) if the support of M_0 is a subset of the support of M_E, and moreover each function and

[145] *PdA* App. 461, *PoA* 456.

[146] Loc. cit. (my emphasys).

[147] *PdA* App. 462, *PoA* 457.

relation of M_0 is the restriction *to the support of* M_0 of the corresponding function and relation in M_E. Unfortunately, this doesn't seem to be the right choice: if M_0 is a *substructure* of M_E, in general only positive atomic assertions concerning elements of the narrower support 'move' from the extension to the substructure. Indeed, assertions like Husserl's example above, 'there are some A', may well be satisfied in M_E, though *not* in M_0, by elements of M_0. A more adequate candidate is surely the notion of *elementary extension*, where a structure M_E is said to be an elementary extension of a structure M_0 of the same type if and only if M_0 is a substructure of M_E and moreover, for every formula $\mathrm{A}(x_1, \ldots, x_n)$ in the first-order language of these structures and every n-tuple $a_1, \ldots, a_n$ of elements of M_0, it holds that $a_1, \ldots, a_n$ satisfy A in M_0 if and only if $a_1, \ldots, a_n$ satisfy A in M_E. However, it is doubtful that when speaking of 'laws (*Gesetze*) that are valid for M_0 (resp. M_E)' Husserl is really thinking of the totality of what we now call elementary conditions.

3.9 The Concept of '*Mathematical Manifolds*'

In the Schuhmann-and-Schuhmann edition of the *Doppelvortrag*, we find some annotations concerning the 'concept of mathematical manifolds' that, in the structure assigned to the text of the *Doppelvortrag* by the editors, should coincide with what Husserl presumably said at the beginning of the second *Vortrag* presented to the *Göttinger Mathematischen Gesellschaft*, on December 10, 1901. Here, too, the text is extremely fragmentary: rather than giving a unitary and systematic treatment of its topic, it appears more like a series of remarks aimed at distinguishing the concept of *mathematical manifold* from that of *definite manifold.*

We can, nevertheless, gather from it evidence for our claim that there are kinds of definite manifold having a certain particular property which makes it possible to give an *arithmetical interpretation* of them, that is to give an arithmetical interpretation of the basic concepts of those manifolds in such a way that the axioms of the corresponding theory become, under this interpretation, arithmetical truths.[148]

To start with, Husserl advances the question whether the concept of a mathematical manifold and that of a definite manifold are equivalent concepts, and he considers some examples of definite manifolds that cannot be mathematical manifolds. Consider for instance a theory regarding a numerical language, containing for example the operations $+$, $\times$, etc. and also the order relation ($\leq$). Let the axioms of the theory be, for example, reflexivity, antisymmetry, transitivity, linearity of the relation $\leq$ and, furthermore, certain principles that connect the operations with the order relation. Let us add to this the axiom: "By such and such axioms the field is determined, and no others are valid for it."[149] Is such a theory definite? Trivially,

[148]On this topic see Section 10 below.

[149]*PdA* App. 442; *PoA* 429.

yes, because we assumed the *axiom of closure*. But it is not a mathematical manifold.

This example should help to understand the following remarks:

> (a) A definite manifold is ruled out by the inessential closure axiom. (b) Can a purely algebraic manifold, which defines no individual of the field whatever [there are no existential axioms, the axioms only define properties of the order relation and connect the order relation with the operations] – can such a manifold have the character of a definite manifold? One can certainly say: If only one operation [e.g. addition] is defined, and if I know that in the deductive sphere in which we are moving only the most general of formulae hold true .... then the associative and commutative laws form a definite combination (*Verbindung*) ... That is, any sentence which contains only the " +," and regardless of how I derived it, is decided as to truth and falsity. Likewise, the well-known laws of addition and multiplication are definite in this sense, under the presupposition of the above supplementary axiom. But it does require precisely the supplementary axiom, and without this that would not hold true.[150]

After having excluded the two cases that we just examined and before considering the philosophically more relevant case of mathematical manifolds, Husserl turns again to the following problem: in what measure is an axiom-system still definite when it is obtained from a definite axiom-system by leaving out one or more axiomatic conditions? This consideration is developed with respect to the structure rather than to the axiom-system. Here, whenever Husserl speaks of 'axiom-system', he means a system of operations. "We now take up operation systems [*Operationssysteme*] which do not exclude the introduction of individuals that give rise to operative results."[151] Take for the moment systems of operations to be essentially *equational* theories in which the axioms express the rules of calculation and the properties (e.g., commutativity or associativity) of certain operations. In a system of operations the "objects" are determined only through operations and are taken as 0-ary operations.[152] According to the partition of the objects of the field which exist on the basis of the axioms, we distinguish two cases:

(i) "Indeterminate objects, that is, objects that do not, by means of the axioms, receive a characterization that turns them into given objects for us",[153] and
(ii) "Determinate objects, that is, individuals defined with their operational axiomatic properties."[154]

Objects of kind (i) would be characterized by the axioms if one or more axiomatic conditions had not been dropped. In Husserl's words, they are not only indeterminate objects "for us," like for instance the unknown variables of an equation of which we do not bother to find the root. These latter kinds of indeterminate objects are fully

150 Sch&Sch 99–100; *PoA* 422–423.

151 Sch&Sch 100; *PoA* 423.

152 See next section for a detailed analysis of this notion.

153 Loc. cit.

154 Loc. cit.

entitled members of the objects of the field.[155] Objects of kind (i) are rather indeterminate objects "*apriori and objectively*", and this depends on the nature of the axiom-system.[156] They do not appear among the 0-ary operations that constitute the determinate objects of the field, and they are never "results" of operations admitted in the field. It is clear, however, that if we add the missing axiomatic conditions they are immediately transformed into determinate objects. "A system of axioms ... to which it is possible to add independent axioms (and which therefore leaves more than one possibility open) is called *disjunctive*."[157] The question is then "whether there can still be individuals apart from these – thus, individuals which can undergo no operational determination and which are absent from the individuals demarcated".[158] Is the structure determined by the restricted axiomatic conditions such that it is no longer possible to add any new object to the domain of the structure? This is the case, for instance, when we consider arithmetic and leave out the axioms that characterize the relation of order. Is such an axiom-system still definite if it is obtained from a definite axiom-system by dropping some axiomatic conditions? In this case, Husserl maintains, we do not have a general rule for determining whether the system of operations is definite or not, hence it has to be established case by case. "It would be definite if, for the demarcated sphere of existence, for the given individuals, and for the individuals not given, no further new axiom were possible."[159]

Finally, the thesis emerges according to which a mathematical manifold is an inductively generated structure. Husserl calls mathematical manifolds also "constructible manifolds (*konstruierbare Mannigfaltigkeiten*)",[160] thereby emphasizing that they are generated by certain modes of construction starting from a set of a given objects. He explicitly allows not only for the case in which the initial elements are fixed, but also for the case in which they are not and where only the procedures of generation are fixed. A system of axioms that defines a mathematical or constructible manifold can

(i) "Either include in its definitions the existence of determinate objects, so that by the univocal forms [*Gebilde*] of operation ever new elements are determined which can then be regarded as given..."
(ii) "Or it can be the case that *only by the arbitrary assumption of a finite number of determinate elements* all others are univocally determinable, as a totality of

[155]"As to the former, they are not merely indeterminate *for us*: Such are also the many generally defined objects of the second field, with respect to the fact that we cannot subjectively work out the general definition, cannot resolve the problematic objects into the determinate objects – and in that sense cannot reduce them to known and given objects" (loc. cit.).

[156]Loc. cit.

[157]I borrow this definition from Hartimo 2007, 296 fn. 29.

[158]Sch&Sch 101; *PoA* 424.

[159]Loc. cit.

[160]*PdA* App. 452; *PoA* 433; Sch&Sch 107.

the possible operational forms, from those determinate elements [arbitrarily assumed]."[161]

In (i) both the set of the initial objects and the generative procedures are fixed; in (ii) the axioms only determine the possible ways to generate new elements starting from those that were arbitrarily chosen as basis. As an example of this second kind of mathematical manifold those structures that are called *free algebras with a given set of generators* might serve.

> Let the system be a mathematical one, That is, any individual existing on the basis of the axioms admits of an operational determination [is generated by operations whose behaviour is fixed in the axioms] and must belong within the sphere of specific operational results (*spezielle Operationsresultate*) [i.e. to the elements generated step by step] (which are obtained on the basis of a certain finite number of objects [the initial elements], whether originally assumed as given in the definition of the manifold [i.e. in the axiom-system] or to be arbitrarily selected and given).[162]

Husserl's idea is that all theories that axiomatize inductively generated manifolds are definite, but not all definite theories axiomatize inductively generated manifolds. Correlatively, on the ontological side, we can say that every mathematical manifold is definite but not every definite manifold is mathematical.

3.10 On the Concept of an Operation System

In Appendix VIII to the *Husserliana* edition of the *Philosophie der Arithmetik*, Husserl characterizes the concept of an *operation system*. We already mentioned this concept when we explained the nature and meaning of mathematical manifolds. Operation systems are mathematical manifolds, but they are not the only ones. For mathematical manifolds are all manifolds that can be arithmetically interpreted, as we shall see in the next section. An operation system is, according to Husserl, a theory on a domain of species. Its peculiarity is that its elements are considered *only insofar as they stand in certain relations*: "Strictly speaking, we have not determined the objects, but rather the system of relations, the system of combinations and relations, and *as* a system that established uniqueness for every term of a relation (*Beziehungspunkt*)."[163] In certain mathematical investigations elements are of interest only as *Beziehungspunkte* of relations. So the difference between a manifold *tout court* and a system of operations is that in the former the axioms determine relations and compositions between elements that, in general, are taken to be *individually different*, whereas in the latter the axioms determine relations and combinations between objects that are *species*.

---

[161] *PdA* App. 458; *PoA* 453.

[162] Sch&Sch 101.

[163] *PdA* App. 476; *PoA* 480.

> We distinguished a definite operation system from a definite manifold. The distinction consists merely in the fact that in the one case "operations" are defined for a domain of species, in the other case relation and relational networks are defined for a domain of elements. We call a domain of species that is defined in a purely formal manner by means of laws of operations (axioms of operation in general) an "operational system".[164]

Husserl takes the elements of an operation system, that is numbers, to be the *infimae species* in the hierarchy of genera and species that obtains among operations in an operation system ("Numbers are the lowest species of operations in a system of operations".[165]) They still admit only a kind of "material fulfillment", which presumably means that numbers are obtained *via* abstraction from sets of concrete objects.[166]

> In arithmetic the a, b, etc. are themselves operations, and 1 is the identity element of multiplication and 0 the identity element of addition. Therefore all the *Operationselemente* themselves have there an operational significance.[167]

> Each letter is itself an indicator of an operation, in such a way, namely, that it represents an object as produced by means of a certain operation type ... $a = b$ says: The objects generated by means of the operation a and the operation b are the same...[168]

As we saw above, in the nowadays current presentation of formal languages one can choose to treat individual constants as functions letters, that is as symbols for 0-ary operations.

In an operation system, compositions between numbers are operations or species of a higher level, that have the job of combining numbers; operations of operations are in turn species of a still higher level, which combine operations. "We have a sphere of species. These species determine new species in virtue of certain combinations existing between them, and between these combinations relations again obtain, and these relations can then in turn serve to determine species."[169]

Examples of operation systems are for Husserl all arithmetics, including the arithmetic of 'types of ordering' (order-types). Thus, types of ordering are *species*, too.

> If a is a type of ordering (*Reihentypus*) (Cantor's order type [*Cantorscher Ordnungstypus*]) and b is a type of ordering (the same or conceptually different), then they determine a new type of ordering, but they do this in the following way: Any ordering of type a can (in the concrete domain of sequences concerned) be combined with a sequence of type b, in such a way that the end point of the one is the initial point of the other. And thereby a sequence, is

[164] *PdA* App. 474; *PoA* 477.

[165] *PdA* App. 480; *PoA* 484.

[166] In the critical annotations to this Appendix we find the following lines cancelled by Husserl with pencil: "The determinate forms of operations (the 'numbers') allow only for a kind of material fulfillment, namely the one according to which the kind of relations, and thereby the genus of the objects upon which the relations are grounded, are materially determined" (*PdA* App. 560–561; not translated in *PoA*).

[167] *PdA* App. 481; *PoA* 485.

[168] *PdA* App. 483; *PoA* 486.

[169] *PdA* App. 481; *PoA* 485.

always determined between the initial point of the first and the end point of the second. This sequence has a type c: $a + b = c$.[170]

This example of type of ordering is interesting because Husserl uses it in his attempt at clarifying what he means, more generally, by *operation* – say, for the sake of simplicity, a binary operation F – *defined on the species* belonging to a certain genus G. The idea seems to be the following: F should be defined "by lifting to the system of species" an operation f on the objects (members of the species) having the property of being *compatible* with respect to the partition into species of the domain of the objects. That is, an f such that for each species a and b and for each pair of objects x, x' of the species a and y, y' of the species b, it holds that the objects $f(x,y)$ and $f(x', y')$ belong to the same species.

Any two individuals . . . of the species a and b (arbitrarily selected form the sphere, where a and b may be of different species or of the same) determine . . . a new object that in turn falls under a species of the genus G, *and the species of the object thus determined must be unambiguously determined*: $a + b = c$.[171]

This clarification is highly significant. The idea behind it is the idea that underlies a well-known abstract construction from *universal algebra*, namely the one that gives rise to a *quotient structure* of a given algebraic structure A by means of an equivalence relation that is also a *congruence* with respect to the operations of A. In other words, Husserl's species correspond to the *equivalence classes* with respect to a certain congruence on the domain of the objects. Recall that a *congruence* over an algebraic structure $A = \langle \mathrm{A}, f, \ldots \rangle$ is a binary relation R on the set A (the support of the structure A) satisfying the following conditions:

(i) R is an *equivalence relation*, that is, it is reflexive, symmetrical and transitive
(ii) R is *compatible* with respect to the operations of A, that is, it satisfies, for each n-ary operation f of A and each $x_1, \ldots x_n, y_1, \ldots, y_n \in A$:

$$R(x_1, y_1) \wedge \cdots \wedge R(x_n, y_n) \rightarrow R(f(x_1, \cdots, x_n), f(y_1, \cdots, y_n)).$$

The quotient of A *modulo* a congruence R is the structure $A/R = \langle \mathrm{A}/R, f_R, \ldots \rangle$ where A/R is the set of the *equivalence classes* $[x]_R =_{\mathrm{df}} \{y \in \mathrm{A} \mid R(x,y)\}$, for $x \in A$, and $f_R([x_1]_R, \ldots, [x_n]_R) =_{df} [f(x_1, \ldots, x_n)]_R$ for each operation f of A.

Finally, Husserl considers the *logical calculus of classes* (*der logische Kalkül der Prädikate*). He maintains that, in this particular case, we are dealing with predicates and modes of combining predicates to obtain new predicates and, consequently, that the letters a, b, ... do not stand for species or, in other words, do not vary on *species* but simply on *classes* or *predicates*. Thus the logical calculus of classes seems not to be an 'operation system' in the sense explained above and, indeed, union and intersection of classes cannot be obtained by "lifting" operations

170 *PdA* App. 483–484; *PoA* 487–488.

171 *PdA* App. 484; *PoA* 488 (my emphasis).

on individuals. However, a few pages later, Husserl argues that the logical calculus is an arithmetic, too; for all "indeterminates (*Unbestimmte*)"[172] of the domain are determinable as True or False, as 1 or 0. Apparently, he is thinking here of the other possible interpretation of the logical calculus: not as a calculus of classes, but as a *calculus of propositions*: "It is not a question of generating from predicates further predicates, as if the calculus were to determine predicates from predicates, but rather it aims to derive true [propositions] from true [propositions], and so in general to draw conclusions concerning truth and falsehood."[173]

Under this second reading, '1' and '0' are no longer the total class and the empty class, but the *species* of all *truths* (of all true propositions), resp. the *species* of all *falsehoods* (of all false propositions). "Thus, as the sphere of the calculus we have the truth-values of propositions . . . and the possible combinations which such truth-values permit."[174] And indeed, the Boolean operations on truth-values can be thought of as deriving (by 'lifting') from composition operations (the connectives) *on propositions* which, insofar as they are *truth-functional* – as one would put it today – are *compatible* with those species '1' and '0'.

Thus the logical calculus, too, is (or can be seen) as an arithmetic, and is 'defined' in the sense that "each letter symbol is either = 0 or = 1, and consequently it is a priori determined, for every relation presenting itself as formula, whether it is satisfied or not. It is, in general satisfied if it, in general, yields 0 = 0 or 1 = 1, otherwise it is false ... The concept of arithmetic must therefore be so broadly conceived that it also includes this case".[175]

3.11 Arithmetizability of a Manifold

All number systems are, as we saw, operation systems. Now, in Husserl the idea is present that each discipline, at least each exact discipline, can be arithmetized. As a theoretical model Husserl follows here Hilbert's arithmetization of geometry,[176] but, of course, all purely mathematical domains such as, for example, the theory of surfaces, the theory of Galois equations, etc. can be arithmetically interpreted. Indeed, it is a fact that this translation is always possible: To each (indeterminate) object of the theory a numerical object is co-ordinated, to each relation of the theory (as e.g. in the case of geometry, "is upon", "lies between") a relation between numerical objects. Each axiom of the theory of the arithmetized theory becomes a

---

[172] *PdA* App. 487; *PoA* 491.

[173] Loc. cit.

[174] Loc. cit. The Fregean sound of this is unmistakable: Frege calls derivations in propositional logic "calculation with truth-values (*Rechnen mit Wahrheitswerten*)" (Frege 1976, 122). But the Boolean framework is entirely antagonistic to Frege: see above App. 5, §3.

[175] *PdA* App. 487; *PoA* 491–492.

[176] Hilbert's *Foundations of Geometry* has been published in Hilbert 1899.

true arithmetical proposition, each theorem a consequence within the theory that interprets it arithmetically.

Husserl thinks that in all theories that axiomatize a definite manifold an arithmetic can be established.

> In geometry [e.g.] an arithmetic can be established, and the axioms can serve to characterize the forms of relation and the modes of determination of the elements. On the basis of the axioms the forms determination are defined, and then the arithmetic of the modes of determination is developed – that arithmetic then serves to determine every element by reference to one that is already given and then to obtain the forms of relation, the relational networks, which we call formations [*Gebilde*], on the basis of those modes of determination: equations in geometry.[177]

His idea is, in particular, that "to any definite manifold a number system can be coordinated that governs all of its relations".[178] Thus the conception of arithmetizability of a manifold can be read as a guide that shows how Leibniz's idea of abstract mathematics as *mathesis universalis* has to be realized. For, on the one hand, each theory is a conceptual framework (*Fachwerk der Begriffe*), as Hilbert once put it, that makes possible the organization of complexes of facts,[179] and, on the other hand, each theory can be interpreted in terms of numbers.

3.12 Husserl's Reappraisal of His Early Theory of Definite Manifolds

In his *Formal and Transcendental Logic* (1929, henceforth: *FTL*) Husserl reconsiders his introduction of the concept of manifold in the *Prolegomena* and in the *Doppelvortrag*.[180] We shall try to provide the conceptual framework for an extensional characterization of Husserl's theory of manifolds (to be set up in the next two sections). In this framework it will be possible to throw light on the conception of the property ascribed by definiteness as categoricity.[181] From the perspective of *FTL* this characterization seems to be very plausible. So it is necessary to return to this topic once again. It should be stated at the outset that Husserl in *FTL* no longer mentions the distinction between absolute definiteness and relative definiteness. Our contention can be formulated as follow: it is true that in *FTL* Husserl believes that definite axiom-systems identify classes of models that are structurally very similar to each other. However, it clearly emerges that his *desideratum* is that to each definite theory there corresponds a definite manifold and *vice versa*. This correspondence does not hold if we interpret the definiteness of an axiom-system as categoricity. So, let us try to find a way out by choosing a weaker notion than

---

[177] *PdA* App. 482; *PoA* 486.

[178] *PdA* App. 475; *PoA* 477.

[179] Hilbert 1918. Cp. Cavaillès 1938, 84 ff.; Casari 2000.

[180] *FTL* 82–85.

[181] Hartimo 2007.

categoricity, for example syntactic completeness, to interpret Husserl's concept of definiteness. Syntactic completeness does not imply categoricity. But a formal counterpart can be given for the structural similarity of the classes of models of a syntactically complete axiom-system. Does the wanted correspondence hold in this case? No, even in this case it does not hold. We didn't really get very far ... We shall now try to reconstruct Husserl's argument step by step, present Null and Simons'[182] attempt to follow that way out and explain why it fails. Thereby we will review an interesting mathematical reconstruction of the theory of manifolds, which illuminates in various respects the conceptual complexity of Husserlian formal logic. In this context we will also make use of syntactical and semantical concepts that in 1929 had not yet received an adequate formulation, although they were present in the mathematical thought of the time as presuppositions or as problems.

In Chapter 3 of *FTL* we firstly come across the metamathematical distinction between a theory as system of axioms and a theory as the corresponding class of models.[183] Husserl writes:

> Since the concept of a theory ... should be understood in the emphatic (*prägnant*) sense ..., that is to say, as a systematic connection of propositions in the form of a systematically unitary deduction, a beginning was found here for a theory of deductive systems ... considered as theoretical *wholes*... As the concept of an objectual totality there appears here what mathematics, without any explication of its sense, understands by *manifold*. It is the form-concept of the realm of objects of a deductive science, where this is conceived as a systematic or total unity of the theory.[184]

We will treat a manifold as the class of all models of an axiom-system on a certain language, and we shall assume that the models of a theory are closed with respect to certain operations on structures: if we perform certain operations on the models of the theory what we obtain are still models of the theory.

Husserl then explores the possibility to formalize a theory. From his observations on formal theories obtained by way of formalization starting from the geometry of our concrete space, we can conjecture that Husserl is trying to characterize the property of isomorphism. He writes:

> The transition to form ... yields the form-idea of any manifold whatever that, conceived as subject to an axiom-system with the form derived from the Euclidean axiom-system by formalization, could be completely explained nomologically, and indeed in a deductive theory that would be (as I used to express it in my Göttingen lectures) 'equiform' to geometry. .... Of course all the concretely exhibited material manifolds that are subject to axiom-systems which, on being formalized turn out to be equiform, have the same deductive science-form in common; they are equiform precisely in relation to this form.[185]

---

[182]Null & Simons 1982.

[183]Null & Simons 1982, 448–449.

[184]*FTL* 78–79.

[185]*FTL* 83.

The term "equiform" makes us think that Husserl conceives the models belonging to such manifold as isomorphic.[186] However, as will soon turn out, Husserl seems to think not only that to each complete system of axioms there corresponds a definite manifold but also *vice versa*. Therefore it is useful to choose a weaker notion than isomorphism, namely second order equivalence, to characterize the property of definiteness of a manifold. Husserl writes:

> When I moved from reflections about the peculiarity of a nomological field to formalization in general, the essential feature of a *manifold form in the emphatic sense*... became apparent. It is defined not by just any formal axiom-system but by a '*complete*' one. Reduced to the precise form of the concept of definite manifold this implies:
>
> The axiom-system formally defining such a manifold is distinguished by the fact that every proposition ... that can be constructed from the concepts ... occurring in that system ... is either 'true' or 'false' ... : *tertium non datur*.[187]

The previous quotation suggests that Husserl not only thinks that to each finitely axiomatized and syntactically complete theory there corresponds a definite manifold but also that to each definite manifold there corresponds a finite syntactically complete system of axioms, in other words, that for each definite manifold it is possible to find a finite number of axioms that allow to decide deductively each formula of the pertinent language.

Like other logicians of the time Husserl tended to use second-order logic and to consider it as complete even though there was no proof of this supposition. Gödel's incompleteness theorem (1931) had not yet been proved, and in order to axiomatize theories as Analysis and Euclidean and Non-Euclidean geometry it was necessary to formally express assertions which refer to all possible subsets of the intended domain. In the formalizations of such theories one pointed to categoricity: the axioms of the theory should identify only one model up to isomorphism.

In the case of theories equiform to geometry, as Husserl calls them, a categorical formalization at the first order is not possible; for the Löwenheim-Skolem-Tarski Theorem shows that if a first-order theory is consistent and has an infinite model then it has models of arbitrary infinite cardinality. A necessary condition for two models to be isomorphic is that their domains are in one-to-one correspondence, hence no first-order consistent theory (having infinite models) is categorical. In order to have categoricity one has to make use of formal languages expressively stronger than first-order languages and allow for quantification over function variables and predicate variables of the language. The models of the theory are taken to be on maximal universes, as one would put it today. In such universes predicate variables vary on all possible properties, that is, extensionally, on all possible subsets of the domain of the individuals.

However, second-order logic is not complete (a consequence of Gödel's incompleteness theorem): there is no system of axioms and inferences rules that is *effective* and characterizes a syntactical concept of derivability that is equivalent

[186]Cp. Null and Simons 450.

[187]*FTL* 100.

to the semantical concept of second-order logical consequence with respect to maximal universes.

3.13 Formal Aspects of the Theory of Manifolds

If one wants to tackle the more proper formal aspects of Husserl's theory of manifolds, the reconstruction given by Null and Simons turns out to be very helpful. The authors take manifolds to be certain well-defined classes of relational structures. They provide for a formal mathematical counterpart of Husserl's theory of manifold that modifies Husserl's notion only when this is made necessary by later results in logic (e.g. incompleteness of second-order logic).

Let us review the authors' choice of primitive logical operators and symbolic notation. The set-theoretic metatheory employed in the paper follows closely those of Gödel, von Neumann, Bernays and Morse: as primitive concepts we find those of *class* (intended as extension of a predicate) and *element or member* of a class. All other concepts of class theory and set theory are defined in terms of these primitives.

Definition. *A class is a set if it is element of some other class, otherwise it is a proper class.*

The symbols $\cap$ and $\cup$ denote the operations of *intersection* and *union* among classes:

1. $A \cap B = [x \mid x \in A \wedge x \in B]$;
2. $A \cup B = [x \mid x \in A \vee x \in B]$.

A is a subset of B, in symbols $A \subseteq B$, iff each member of A is a member of B. A is a proper subset of B ($A \subset B$) iff $A \subseteq B$ but not $B \subseteq A$. A n-ary relation R (on C) is a class of ordered n-tuples (of elements of C). The symbol '$\emptyset$' denotes the empty set.

An equivalence relation on a class C is a binary relation R on C which satisfies, for each element x, y, z of C, the conditions:

a. Reflexivity: $< x, x > \in R$
b. Symmetry: if $< x, y > \in R$, then $< y, x > \in R$
c. Transitivity: if $< x, y > \in R$ and $< y, z > \in R$, then $< x, z > \in R$

The formal languages employed are *monadic second-order languages with identity*. These languages (from now on denoted by L, L', ...) are based on an alphabet which contains, besides the usual auxiliary symbols:

1. the usual logical constants (connectives and quantifiers), and identity
2. countably many individual variables
3. countably many monadic predicate variables: $X_1, X_2, X_3, \ldots$
4. a (possibly infinite) number of predicate constants $P_1, P_2, \ldots$ (of arity greater than zero)

The symbols listed in (1–3) are common to all such formal languages, whereas those in (4) vary from one language to another. Terms and formulae of L are built up following the usual recursive procedure as in (purely predicative) monadic first-order languages, with the fundamental difference that in the construction of the formulae we allow for quantification not only of individual variables but also of monadic predicate variables.

Let us now recall the definition of *model* for a language L. Given a language L with the predicate constants $P_1, P_2, \ldots$, a model M for that language is a relational structure made up of a domain D and certain relations, defined on that domain, which are in one-to-one correspondence with the predicate constants of L and taken to be the meanings of the latter:

$$M = < D, R_1, R_2 \ldots >,$$

where P_i and R_i (the meaning of P_i in M) have the same arity.

Following Tarskian semantics, the concepts of *satisfaction* and *truth* of a L-formula in a L-model are defined as usual. As the languages in question are monadic second-order languages, the quantificational interpretation of predicate variables is the *maximal* one, that is, the field of variation of monadic predicate variables is the set of *all* subsets of the domain D of M.

If α is a formula, we write '$M \models \alpha$' to say that α is true in M.

Definition. *Given an L-model M and a set of axioms* Σ *in the language L, M is a model of* Σ *iff* $M \models \alpha$ *for each* α *belonging to* Σ.

Let us now give the definition of deductive system and formal proof.

Definition. *A deductive system S for a language L is an effective system of axioms and/or inference rules.*

Definition. *A S-derivation of a formula* α *from a set of formulae* Σ *is a finite sequence of formulae each of which is either an axiom of S or a formula in* Σ *or it is obtained from previous formulae in the sequence by way of an application of an inference rule of S. A S-theorem of* Σ *is a formula of the language of* Σ *that is S-derivable from* Σ.

The properties of correctness and general completeness of a deductive system S are specified in the usual way:

1. *Correctness*: for each Σ and for each formula α (of the language of Σ), if α is S-derivable from Σ then α is true in each model of Σ.
2. *Completeness*: for each Σ and for each α (of the language of Σ), if α is true in each model of Σ then α is S-derivable from Σ.

The soundness and completeness theorems hold for first-order predicate logic. Presumably Husserl thought that the two theorems also hold for second-order predicate logic. In what follows, we will refer to an *arbitrary* deductive system S for second-order monadic logic that (i) is sound, and (ii) extends a sound and complete deductive system for first-order logic. The symbol '$\vdash$' denotes the relation of S-derivability.

As we will use the relations of *isomorphism*, *elementary equivalence* and *second-order* (monadic) *equivalence* between models as basic concepts for the development of Husserl's conjectures, let us give their definition and recall three theorems which establish that isomorphism is an equivalence relation and that isomorphism-types and equivalence-types are proper classes.

Definition. *Two models (for the same language L)*

$$M = <D, R_1, \ldots, R_n \ldots> \text{ and } M' = <D', R'_1, \ldots, R'_n \ldots>$$

are isomorphic if there is a function f from D to D' such that

(a) f is a bijection, i.e., f is:

- injective: if $f(x) = f(y)$ then $x = y$ (for $x,y \in D$)
- surjective: for all $y \in D'$ there is an $x \in D$ such that $f(x) = y$

(b) The relations R_i and R'_i have the same arity, and for each n-tuple $x_1, \ldots, x_n \in D$ (where n is the arity of R_i) it holds:

$$R_i (x_1, \ldots, x_n) \text{ iff } R'_i (f(x_1), \ldots, f(x_n)).$$

Intuitively, two models are isomorphic if they are structurally indiscernible, that is, they differ only with respect to the nature of the elements of the two domains. A set of axioms Σ is *categorical* if all its models are isomorphic.

We further distinguish between elementary and monadic second-order equivalence:

Definition. *Two models M and M′, for the same language L, are elementary equivalent (in symbols $M \equiv M'$) if for each closed elementary formula α of the language it holds: $M \models \alpha$ iff $M' \models \alpha$. That is, two models are elementary equivalent if they make true exactly the same elementary formulae.*

Definition. *Two models M and M′, for the same language L, are monadic second-order equivalent ($M \equiv_{m2} M'$) if, for every α, $M \models \alpha$ iff $M_O \models \alpha$.*

Theorem 1. *Isomorphism is an equivalence relation (over the class of models for a given language).*

Given a class C and an equivalence relation R on C, R yields a partition of C into *classes* or *equivalence-types*; that is, the relation R divides all individuals of the class C into non-empty and disjoint classes of elements. Each of these classes contains all and only the elements of C that stand to each other in the equivalence relation R.

Given the class of all models of a certain language L, the isomorphism relation produces equivalence classes that we call *L-isomorphism types.*

Definition. *An L-isomorphism type is the class of all models (of L) that are isomorphic to any model (of L).*

Theorem 2. *Each L-isomorphism type is a proper class.*

Elementary and monadic second-order *equivalence types* are analogously defined.

Theorem 3. *Each (elementary or monadic second-order) equivalence type is a proper class.*

Theorem 3 can be proven by the observation that each equivalence type contains an isomorphism type as subclass. It follows from the above that each theory concerning isomorphism types, equivalence types and manifolds is committed to assign properties and to establish relations on proper classes.

We will now introduce the notions of *manifold, Husserl-definite manifold* and *formal manifold.* We will then present some theorems concerning properties of manifolds. Each manifold will turn out to be a proper class. We will then consider some theorems on the fundamental properties that Husserl seems to ascribe to *definite* manifolds, in particular the theorem which establishes that a manifold determined by a set of axiom is *definite* iff the system of axioms is *syntactically complete*. Of this statement, only one direction holds. For the other direction we have, because of Gödel's incompleteness theorem, a counterexample at disposal that proves that the biconditional does not hold.

Definition. *A manifold* $\boldsymbol{M} = M[\Sigma]$ *is the set of all models of an axiom-system* Σ *on a language L.*

Theorem 4.

(a) Each manifold is a union of isomorphism types.
(b) Each manifold is a proper class.

Definition. *A manifold* $\boldsymbol{M} = M[\Sigma]$ *is Husserl-definite if it is a* $\equiv_{m2}$*-type. In other words if it holds that*

(a) if $M \in M[\Sigma]$ and M' is monadic second-order equivalent to M, then also $M' \in M[\Sigma]$, and
(b) if $M \in M[\Sigma]$ and $M' \in M[\Sigma]$ then M and M' are monadic second-order equivalent

Theorem 5. *Each manifold which is an isomorphism type is Husserl-definite.*

Proof. Let $\mathbf{M} = M[\Sigma]$ be an isomorphism type.

We assume:

(i) $M \in M[\Sigma]$ and (ii) $M' \equiv_{m2} M$

If $\alpha \in \Sigma$, for (i) it holds $M \models \alpha$, therefore for (ii) it holds $M' \models \alpha$; hence M' is a model of Σ and therefore *per definitionem* $M' \in M[\Sigma]$.

We now assume $M, M' \in M[\Sigma]$. $\boldsymbol{M}$ is, according to the hypothesis, an isomorphism type. Now isomorphic models are trivially $\equiv_{m2}$-equivalent. So $M \equiv_{m2} M'$, and M is an Husserl-definite manifold.

Definition. *A manifold* $M[\Sigma]$ *is formal iff it is an isomorphism type, that is, iff all models of* Σ *are isomorphic.*

This means that the axiom set Σ is categorical. It follows from the definitions and Theorem 5 that *some* Husserl-definite manifolds are formal and that each formal manifold is Husserl-definite.

It is now possible to prove that, given a system of axioms Σ that is *syntactically complete* (for each closed formula α of the language it is possible to syntactically decide α on the basis of Σ, that is, either $\Sigma \vdash \alpha$ or $\Sigma \vdash \neg\alpha$ holds), the models of Σ are all monadic second-order equivalent.

Theorem 6. *For each syntactically complete axiom-system Σ, the manifold $\mathbf{M} = M[\Sigma]$ is Husserl-definite.*

Proof. It suffices to show that, under the hypothesis that Σ is syntactically complete and that $M, M' \in M[\Sigma]$, it holds $M \equiv_{m2} M'$. Suppose $M \models \alpha$. Then M is a model of Σ, hence for the property of soundness assumed for the deductive system S it cannot be the case that $\Sigma \vdash \neg\alpha$. Hence, for the hypothesis of syntactic completeness of Σ, it must be the case that $\Sigma \vdash \alpha$. But M' is also a model of Σ and hence, still for the property of soundness, $M' \models \alpha$. In an analogous way we prove that if $M' \models \alpha$ then $M \models \alpha$. Thus, the conclusion is that $M \equiv_{m2} M'$.

Husserl seems to maintain that also the other direction of Theorem 6 holds: syntactic completeness would then represent a necessary and sufficient condition of the definiteness of a manifold. This, however, is false, as can be shown by applying Gödel's incompleteness theorem.

The following Lemma holds:

Lemma 7. *For each set of axioms Σ, if Σ is categorical, then $M[\Sigma]$ is a Husserl-definite manifold, or rather (by Theorem 5) a formal manifold.*

We can now prove the following theorem.

Theorem 8. *There is an infinite system of axioms Σ, that is categorical and hence, because of Lemma 7, such that the manifold $M[\Sigma]$ (the class of models of Σ) is Husserl-definite. But that system is not syntactically complete.*

Proof. Let PA_2 be the set of the following five axioms[188]:

N1.	$\forall x[\neg s(x) = 0]$
N2.	$\forall x \forall y[s(x) = s(y) \rightarrow x = y]$
N3.	$\forall X[X(0) \wedge \forall y(X(y) \rightarrow X(s(y))) \rightarrow \forall x X(x)]$
N4.	$\forall x \forall y[(x + 0 = x) \wedge (x + s(y) = s(x + y))]$
N5.	$\forall x \forall y[(x \cdot 0 = 0) \wedge (x \cdot s(y) = x + (x \cdot y))]$

The language of PA_2 contains the individual constant '0' for the number zero, the unary function letter 's' for the successor operation and the function letters '+' and '·' for addition and multiplication. Thus it is not purely predicative. Let then Σ be a finite set of axioms that is semantically equivalent to PA_2 and is obtained by

[188]The axiomatization of PA_2 in Null & Simons 1982, 452–453, is slightly different from ours, although trivially equivalent to it. This is due to the fact that they do not take the individual constant '0' as primitive, but replace it by a primitive binary predicate constant '<' (expressing the standard order relation).

conveniently recasting the language of PA_2 as a second-order purely predicative monadic language (which is always possible).

Because of the theorem of categoricity of arithmetic (Dedekind 1888) all maximal models of PA_2, and thus of Σ, are isomorphic; the manifold $M[\Sigma]$ is therefore formal and *a fortiori* Husserl-definite. However, due to Gödel's incompleteness theorem, there is no deductive system S (in the sense of the definition previously given) with respect to which the axiom-system Σ is syntactically complete.

The following theorem establishes a *weaker* correspondence between the definiteness of a manifold and the completeness of some axiom-system that determines it.

Theorem 9. *For every manifold* $\boldsymbol{M}$*, if* $\boldsymbol{M}$ *is Husserl-definite, then there is a possibly infinite set of axioms* Σ' *such that* $\boldsymbol{M} = M[\Sigma']$ *and* Σ' *is syntactically complete.*

To prove Theorem 9 it suffices to observe that each manifold $\boldsymbol{M}$ is conceived as a manifold determined by an axiom-system Σ (i.e. $\boldsymbol{M} = M[\Sigma]$). If $\boldsymbol{M}$ is Husserl-definite and Σ is not syntactically complete, it can always be extended to a system of axioms Σ' that is syntactically complete and that determines the same manifold: it suffices to take as Σ' the set of *all* closed formulae that are true in some model of Σ (and therefore also in all models of Σ, because $\boldsymbol{M}$ is a $\equiv_{m2}$-type). This system of axioms Σ' is, however, *infinite* (and non-effective), while for Husserl axiom-systems ought to be *finite*. Husserl never considered the notion of an infinite axiom set, which arguably does not fit well with his philosophical goals.

This formal reconstruction of Husserl's theory of manifolds is based upon second-order object languages. If we restrict ourselves to first-order languages, it can be proved (by applying the *completeness* theorem for first-order logic) that a system of axioms is syntactically complete iff the manifold is definite.

A *first-order Husserl-definite manifold* is a $\equiv$ -type. A system of axioms Σ on a first-order language (without quantification on predicate variables) is syntactically complete iff for every proposition α of the language of Σ, either α or $\neg\alpha$ is a theorem of Σ.

Theorem 10. *For every first-order axiom-system* Σ*, the manifold* $M[\Sigma]$ *is a first-order Husserl-definite manifold iff* Σ *is a first-order syntactically complete axiom-system.*

Proof. From right to left the theorem is to be proved in a way wholly analogous to Theorem 6. From left to right: Let $M[\Sigma]$ be a first-order Husserl-definite manifold. Let α be a first-order closed formula, such that $\Sigma \vdash \alpha$ does *not* hold. Because of the general completeness theorem for first-order logic, there is a model M of Σ (hence, an $M \in M[\Sigma]$) such that $M \models \alpha$ does *not* hold, that is $M \models \neg\alpha$. However, as $M[\Sigma]$ is assumed to be a first-order definite manifold, each model $M' \in M[\Sigma]$ is elementarily equivalent to M, and thus makes true all propositions that M makes true. In particular, as $M \models \neg\alpha$, $\neg\alpha$ will be true in every model of Σ. In virtue of the general completeness theorem this implies that $\Sigma \vdash \neg\alpha$. Thus we have proved that if *not* $\Sigma \vdash \alpha$ then $\Sigma \vdash \neg\alpha$. This is equivalent to saying that Σ is syntactically complete at the first-order.

3.14 Ways of Generalization

In order to introduce the concept of *generalization*, understood as a relation between manifolds, we first give the definitions of *expansion* and *restriction* of a model and impose some constraints on models.

Definition. *An expansion of a model M is a model M′ whose domain is the same as that of M, and whose relations are those of M, plus possibly some others.*

With respect to the language, this means to add new predicate constants to which there correspond, on the semantical level, new relations.

Definition *M is a restriction of M′ iff M′ is an expansion of M.*

From now on we will consider only models that satisfy the following additional constraints:

(1) The domain of *M* has at least two elements
(2) For every *n*, the intersection of a finite number of *n*-ary relations of *M* is non-empty

Null and Simons consider three different kinds of generalization for Husserl-definite manifolds: generalization *by weakening axioms*, generalization *by removals, and* generalization *tout court.*

3.14.1 Generalization by Weakening Axioms

Let $\boldsymbol{M_1}$ and $\boldsymbol{M_2}$ be two manifolds. By definition of manifold they will be of the form $M[\Sigma_1]$ and $M[\Sigma_2]$ for certain sets of formulae Σ_1 and Σ_2.

Definition. *$\boldsymbol{M_1}$ is a generalization of $\boldsymbol{M_2}$ by weakening axioms iff $\boldsymbol{M_2} \subset \boldsymbol{M_1}$.*

In other words, all models of Σ_2 are also models of Σ_1, but not *vice versa.*

We recall that there is an inverse correspondence between the axioms of a theory and the corresponding models:

$$\text{if } \Sigma_1 \subseteq \Sigma_2 \text{ then } M[\Sigma_2] \subseteq M[\Sigma_1],$$

that is to say, given two axioms systems one of which is included in the other, the corresponding sets of models stand in the inverse correspondence. Intuitively, the fewer constraints one imposes, the more structures one characterizes. In the case of generalization by weakening axioms, it is possible to prove that

(i) if $\alpha \in \Sigma_1$ then $\Sigma_2 \models \alpha$
(ii) there is at least one $\alpha \in \Sigma_2$ such that *not* $\Sigma_1 \models \alpha$

The system of axioms relative to the manifold that generalizes turns out to be weaker with respect to the one characterizing the manifold on which the generalization is performed: Σ_1 is included among the logical consequences of

Σ_2 (the axioms of Σ_1 are not necessarily among the axioms of Σ_2, but they are in any case logical consequences of Σ_2).

Having fixed the meaning of generalization by weakening axioms, the authors prove the following three theorems:

Theorem 16. *For each isomorphism type **I** there is a Husserl-definite manifold **M** such that **I** is a subclass of **M** and each generalization of **M** is obtained by weakening axioms.*

Theorem 17. *Each Husserl-definite manifold can be generalized by weakening axioms in a manifold that is no longer a Husserl-definite manifold.*

Theorem 18. *A manifold is Husserl-definite iff there is no manifold of which it is a generalization by weakening axioms.*

Among manifolds, Husserl-definite manifolds turn out to be the *minimal* elements with respect to the generalization by weakening axioms.

3.14.2 Generalization by Removals

Let $\boldsymbol{M_1} = M[\Sigma_1]$ and $\boldsymbol{M_2} = M[\Sigma_2]$ be given. To have a generalization by removals of $\boldsymbol{M_2}$ by $\boldsymbol{M_1}$ the following constraints are to be imposed:

(1) $\Sigma_1 \subset \Sigma_2$
(2) Each model M of $\boldsymbol{M_2}$ is an expansion (in the sense defined above) of some model M' belonging to the manifold $\boldsymbol{M_1}$

Intuitively, this second kind of generalization of a manifold $\boldsymbol{M_2} = M[\Sigma_2]$ consists, on the one hand, in the removals of some axioms from Σ_2, thereby obtaining a subsystem Σ_1 of it, and, on the other hand, in the removals of certain predicates from the language of Σ_2 (precisely, those predicates that appear in some axioms of Σ_2, but do not appear any more in any axiom of its subsystem Σ_1), and in the removals of the relations corresponding to those predicates from each model of $\boldsymbol{M_2}$. And all this in a way that each model of Σ_2 turns out to be an expansion of some model of Σ_1 by means of the meanings for the predicates that could have gone lost because of the removals.

3.14.3 Generalization "Tout Court"

In conclusion, we define $\boldsymbol{M_1}$ as a generalization *"tout court"* of $\boldsymbol{M_2}$ iff $\boldsymbol{M_1}$ is a generalization by weakening of axioms of $\boldsymbol{M_2}$, or $\boldsymbol{M_1}$ is a generalization by removals of $\boldsymbol{M_2}$, or there is a manifold $\boldsymbol{M_3}$ such that $\boldsymbol{M_3}$ is a generalization of $\boldsymbol{M_2}$ by removals and $\boldsymbol{M_1}$ is a generalization of $\boldsymbol{M_2}$ by weakening the axioms.

Theorem 18 (which refers to the generalization by weakening the axioms) has an analogoue that concerns the more general notion of generalization "tout court".

Before presenting it is necessary to introduce the notion of a model that is a *full expansion.*

Definition. *A model M is a full expansion iff for each n, the set of its n-ary relations is an ultrafilter on the domain of M.* We recall that, given a set A, an ultrafilter U over A is a set of subsets of A that

1. does not contain the empty set $\emptyset$
2. is closed under $\cap$ (set-theoretical intersection)
3. if $X \in U$ and $X \subseteq X'$ then $X' \in U$ (for all $X \subseteq A$)
4. given any subset X of A, either X or its complement $-X$ is in U

It can be proved (respecting the restrictions here imposed on the models) that

(i) no model that is a full expansion has a proper expansion;
(ii) each model is either a full expansion or has an expansion that is a full expansion.

Finally, it is possible to prove:

Theorem 20. *A manifold is a Husserl-definite manifold of full expansions iff there exists no manifold of which it is a generalization tout court.*

Intuitively: there is no manifold of which a Husserl-definite manifold of full expansions is a generalization *tout court*. One can think of a Husserl-definite manifold of full expansions as an atom: it is at the highest level of definiteness.

3.15 Appendix 6: Husserl's Existential Axiomatics

On the following pages we shall focus on the text "*Das Gebiet eines Axiomensystems/Axiomensystem – Operationssystem*" which is "Anhang VIII, Studie I" of the *Husserliana* edition of the *Philosophie der Arithmetik.*[189] Its initial part – acknowledged by Schuhmann & Schuhmann as belonging to the *Doppelvortrag* – is included in their edition under the title "*Transcript from the lecture* (*Abschrift aus dem Vortrag*)".[190] The main point of interest of this text is Husserl's discussion, and acceptance, of some essential aspects of a method that Hilbert and Bernays came to call "existential axiomatics".[191] This connection will open the way for some reflections concerning a hierarchical order of manifolds and the property of definiteness.

---

[189] *PdA* App. 470–474; *PoA* 420–422 & 475–477.

[190] Sch&Sch 98–99; *PoA* 420–422.

[191] Our analysis has profited a lot from Sieg 2002. It was only by reflecting on Sieg's illuminating account of the problems connected with "existential axiomatics" that we (hopefully) found a key for the interpretation of Husserl's talk of existential axioms.

Husserl's goal in our text is to establish "what is to be understood under a field (*Gebiet*) of an axiom-system".[192] So he begins with the question: "The axiom-system 'defines' a field... What does that mean?" Generally, he maintains, "we can say of any consistent (*verträglich*) collection of formal conventions [axioms] ... that it defines a field, a manifold of objects." However, "a special case is to be marked out; the concept of the field of an axiom-system, and the sense of the assertion 'an axiom-system has a field' has to be provided with a richer content."

Strictly speaking, Husserl contends, an axiom-system defines a field just in case it includes "existential axioms (*Existentialaxiome*)".[193] To understand what Husserl means by this phrase, it will turn out to be useful to remember that Dedekind already in 1888 and 1890, in his *Was sind und was sollen die Zahlen?* and in his explanatory letter to Keferstein, explicitly posed questions of existence. In §§ 71 and 73 of his book Dedekind explains that N is a *simply infinite system* just in case there is an injective map ϕ of N in itself together with a privileged element 1 of N which does *not* belong to the codomain of ϕ and is such that its *chain* 1_0 (the intersection of all subsets of N which contain 1 and are closed under ϕ) coincides with N itself. That simply infinite systems exist follows (§ 72) from the main claim of § 66 in which Dedekind proves that "there are infinite systems"[194] by showing that the "totality of all things that can be object of my thought" actually is an infinite system:

> My own realm of thoughts, i.e. the totality S of all things, which can be objects of my thought, is infinite. For if s signifies an element of S, then the thought s' that s can be an object of my thought is itself an element of S.[195]

For Dedekind not only the notion of a simply infinite system is crucial but also its non-emptiness. He returns to this point in a letter to Keferstein:

> After the essential nature of the simply infinite system, whose abstract type is the number sequence N, has been recognized in my analysis ... the question arose: does this system exist at all in the realm of pure thought? Without a logical proof of existence it would always remain doubtful whether the notion of such a system might not perhaps contain internal contradictions. Hence the need for such formal proofs.[196]

So Dedekind clearly recognizes the need for a proof of existence for a system satisfying certain abstract conditions.

In 1904 Hilbert criticized Dedekind's alleged 'logical proof' in a paper *Über die Grundlagen der Logik und der Arithmetik* he read in Heidelberg at the Third International Congress of Mathematicians:

---

[192]*PdA* App. 470; *PoA* 420 (All quotations in the paragraph are from this page).

[193]Loc. cit.

[194]According to Dedekind 1888, §60 a system S is said to be infinite if it similar to (i.e., it can be put into one-to-one correspondence with) a proper part of itself.

[195]Dedekind 1888, 14. Dedekind registers the fact that a similar consideration can be found in §13 of Bolzano's *Paradoxien des Unendlichen.*

[196]Dedekind 1890.

> R. Dedekind clearly recognized the mathematical difficulties one encounters when a foundation is sought for the notion of number: for the first time he offered a construction of the theory of integers, and in fact an extremely sagacious one. However, ... in proving the existence of the infinite he follows a method that ... I cannot recognize as practicable or secure, for it employs the notion of the totality of all objects, which involves an unavoidable contradiction. G. Cantor was aware the contradiction just mentioned, and he expressed this awareness by distinguishing between "consistent" and "inconsistent" sets.[197]

This does not mean that questions of 'existence' in the spirit of Dedekind do not matter to Hilbert. In his paper *Über den Zahlbegriff*,[198] published in 1900 (and presented one year before at a meeting of the *Deutsche Mathematische Vereinigung* in Munich), before presenting the axiomatic conditions for the number system of the reals, Hilbert contrasts 'the formal axiomatic method' with the 'genetic method.' The latter consists, as we saw, in obtaining the reals by stepwise expansion of the system of natural numbers. To this method Hilbert attributes at least a pedagogical or heuristic value. By contrast, he maintains, the construction of geometry is realized by a completely different method: "one begins *by assuming the existence* of all elements (that is one assumes at the beginning three different systems of things: points, lines and planes) and one puts these elements into certain relations to one-another by means of certain axioms, in particular the axioms of connection, order, congruence and continuity".[199] Hilbert asks whether the genetic method really is the only adequate method for the study of the number concept while the axiomatic method is the most adequate method for laying out the foundations of geometry. And his answer is that if one wants to present the number system of the reals adequately and to ensure logically the significance of our knowledge-claims in this area, one ought to choose the axiomatic method.

After these methodological clarifications he goes on to list the axiomatic conditions for the reals.[200] They are introduced by assuming *the existence of a system of things that satisfies certain conditions*: "We think a system of things. Let us call these things numbers and denote them by *a*, *b*, *c* .... We think these numbers in certain mutual relations whose precise and complete description happens by means of the axioms."[201] In his 2002 Sieg rightly stresses that Hilbert follows Dedekind here:

> Thus, as in Dedekind's case, there is an explicit existential assumption that has to be secured or discharged in some way. To emphasize this crucial aspect of Hilberts method, both Hilbert and Bernays called it "existential axiomatics (*existentiale Axiomatik*)".[202]

---

197 Hilbert 1905.

198 Hilbert 1900a.

199 Hilbert 1900a, 181 (my emphasis). Cp. Reid 1970, 68–69.

200 Hilbert 1900a, 181–182.

201 Hilbert 1900a, 181.

202 Sieg 2002, 367–368.

It will help to clarify Husserl's talk of existential axioms if we recall some of Hilbert's axioms for the reals (taken from the first group, the "axioms of collection"[203]):

Axiom I.1. From the number a and from the number b we obtain by means of "addition" a determinate number c; in symbols: $a + b = c$ or $c = a + b$.

Axiom I.3. There is a determinate number – let us denote it by '0' – such that for each a it holds: $a + 0 = a$, and $0 + a = a$.

Axiom I.4. From the number a and from the number b one obtains by means of "multiplication", a determinate number c; in symbols: $a \cdot b = c$ or $c = a \cdot b$.

Axiom I.6. There is a determinate number – let us denote it by '1' – such that for each a it holds: $a \cdot 1 = a$, and $1 \cdot a = a$.

When talking of 'existential axioms' Husserl means axioms given exactly in this way. For him an axiom-system defines a field only if it includes existential axioms of this kind. Thus, if the field under consideration is e.g. the one formally defined by the axioms of arithmetic, one stipulates that "*there is* a composition '+' (which implies that *there are* determinate pairs of elements a, b, which are combinable in the form $a + b$, and 'combinable' means in turn: *there is* in it at least one new element, which equals $a + b$), and for this combination such and such laws are valid."[204]

Husserl's argument in the *Transcript from the Lecture* can be summarized as follows:

1. We think of a system of things.
2. We establish existential axioms that determine which objects belong to the field.
3. We establish axioms that fix the relations and the combinations between the objects of the field.
4. We ask in which sense the specified axioms can be said to univocally identify a certain system.

Among existential axioms Husserl distinguishes those that are 'univocal (*eindeutig*)' from those that are 'determinately equivocal (*bestimmt vieldeutig*)' and those that are 'indeterminately equivocal (*unbestimmt vieldeutig*)'. He writes:

> These existential axioms can be univocal or equivocal, and in the latter case again either determinately or indeterminately equivocal. If we now totally exclude the case of being indeterminately equivocal, then determinate equivocality can be eliminated by the joint force of the axioms, so that we are able to determine univocally ever new elements from given elements (and here that can only mean: elements assumed as given and, as it were, named by proper names) on the basis of the axioms and consequently to regard them likewise as given. Of an axiom-system which delimits in this manner a general sphere of univocally determinate existents and thus contains forms ... of univocal determination of objects from which forms ... ever new elements of a derived kind result: of such an axiom-system we say that it has a domain.[205]

[203]Hilbert 1900a, 181–182.

[204]*PdA* App. 470; *PoA* 420.

[205]*PdA* App. 470–471; *PoA* 420–421.

In this passage Husserl tries to bring two different kinds of problems to light. The first can be classified under the heading 'saturation of an axiom-system', the second is connected with the above mentioned view that "an explicit existential assumption has to be secured and discharged in some way." Let us confront the first problem.

The case of existence axioms that are indeterminately equivocal is excluded. But an ambiguously determinate axiom-system seems to correspond to what Veblen later called a disjunctive system: it can be extended by adding a new *independent* axiom. Thus the role of each further axiom is that of introducing a limitation in such a way that the sphere of ambiguity progressively disappears. In other words, an 'ambiguously determined axiom-system' is such that its saturation is always possible: "whatever is ambiguous . . . must only be ambiguous because of the incompleteness of its determination and must be capable of transformed into something unambiguous. In the manifold or in the axioms nothing can remain that is ambiguous 'in principle'".[206] In Chapter 3 we faced the difficulty of making the notion of saturation precise. Axiom-systems containing existential axioms that are 'determinately equivocal' admit models that are formally/structurally different. They leave open the possibility of bifurcation (*Gabelbarkeit*), as Veblen put it.[207] On the ontological side, each additional axiom does the job of further specializing the corresponding manifold.

> If the definition of a manifold [sc. its axiom-system] does not unambiguously determine its objects in relation to each other, it expresses formal relationships that can not only individually but also specifically belong to [i.e. characterize] manifolds of different forms.[208]

Finally, existential axioms are called *univocal* if they univocally fix (i) a (finite) number of initial objects and (ii) the generative procedures to obtain all objects of the domain starting from the given ones.

> The most restricted case is the one when the domain coalesces to form one single field of operations (*Operationsfeld*): that is, where all existence is enclosed in that which is constructible from a finite number of pre-given objects.[209]

Existential axioms are also univocal when they univocally fix only the generation procedures and leave the initial elements indeterminate. In both cases, 'univocity' coincides with the impossibility of *bifurcation* (*Nichtgabelbarkeit*).[210] In this case, no *further specialization* is possible for the corresponding manifold. The system of axioms for such a manifold "no longer admits of a further determination

---

[206] *PdA* App. 477; *PoA* 480–481.

[207] Cavaillès 1938, 91–92.

[208] *PdA* App. 470; *PoA* 475.

[209] *PdA* App. 471–472; *PoA* 421–422.

[210] Cavaillès 1938, 91–92.

... so that the manifold can still be individualized but is no longer determinable (differentiable) as to its form". [211]

To summarize, given a defining axiom-system or theory, "AX", Husserl distinguishes two cases:

1. The theory AX is consistent but not saturated, i.e., there is at least one proposition A of its language that the system does not decide:

$$\exists A[\textit{not}(\mathrm{AX} \vdash A)\ \&\ \textit{not}(\mathrm{AX} \vdash \neg A)].$$

In this case the theory can be specified further. How can it be expanded in a way that it remains consistent? For every statement A, A represents a genuine possibility to specify the theory if and only if *not*($\mathrm{T} \vdash A$) and $\mathrm{T} + A$ is consistent, i.e. *not* ($\mathrm{T} \vdash A$). So $\mathrm{T} + A$ is a genuine expansion of T iff *not*($\mathrm{T} \vdash A$) and *not*($\mathrm{T} \vdash \neg A$), i.e. *T does not decide A* .

2. The theory AX is saturated and determines completely the corresponding manifold.

In this case – according to Husserl – there still remains the possibility of *expanding the domain* of the theory by adding new objects (and, correspondingly, expanding the language of the theory). Yet he explicitly says that this latter kind of expansion does not constitute a real specification: "But we will not call an expansion of the domain a specification (*Spezialisierung*) in the true sense".[212]

The distinction between 'ambiguously' and 'unambiguously' determinate axiom-systems discussed so far opens the way for establishing a *hierarchy* of genera and species in the realm of (forms of) manifolds and, correlatively, of (forms of) theories.

> The forms of manifold (and, with them, the forms of theory) thus constitute a realm (*Reich*) with tiers of genera and species. The lowest species are, as it were, the individuals of this realm. These "individuals" are the complete (*perfekt*) forms of manifolds.[213]

A manifold is, as we saw, 'completely definite (*perfekt definiert*)'[214] if the corresponding axiom-system is *saturated*. In this case, Husserl claims, the manifold is only materially determinable (*materiell bestimmbar*), i.e. only an 'individualization' of it is still possible.

> A formally defined manifold is completely determined as to its form if nothing more remains open formally [each new axiom would make the theory inconsistent, – the axiom-system is maximal]. In this case the manifold is only materially determinable, its concept is an ultimate specific difference.[215]

---

[211] *PdA* App. 473; *PoA* 476.

[212] *PdA* App. 473; *PoA* 477.

[213] Loc. cit.

[214] Loc. cit.

[215] *PdA* App. 472; *PoA* 475.

Such a manifold can be considered as an atom, an *infima species* in the hierarchy of manifold forms, a manifold whose axiom-system is at the highest degree of saturatedness. In which sense can such a manifold still be expanded?

> Such a manifold still allows for new stipulations of existence and for corresponding rules of operation, but *not for new general law of operation, and above all, not for a new axiom for the old domain.*[216]

That is to say, it is no longer possible to add to the corresponding theory new independent axioms *dealing with the originally intended components*. However, there remains the possibility of expanding the domain underlying the theory T and of adding new operations on the extended domain.

At this point some hints as to a conceptual distinction Kit Fine made in a different context may turn out to be helpful. In his paper 'Our knowledge of mathematical objects'[217] Fine proposes a new approach to the philosophy of mathematics that he calls '*procedural postulationism*'. This account differs from the standard one by the feature that a mathematical domain is not conceived as a model of certain axioms but rather as a structure generated by one or more *procedural postulates*, i.e. procedures for the construction of that domain.

> Just as a computer program specifies a set of instructions that govern the state of a machine, a postulate will specify a set of instructions that govern the composition of the mathematical domain; and just as the instruction specified by a computer program will tell us how to go from one state of the machine to another, so the instructions specified by a postulate will tell us how to go from one 'state' or composition of the mathematical domain to another (one that, in fact, is always an expansion of the initial state).[218]

Although Fine never mentions Husserl in his paper, some of the main features of '*procedural postulationism*' come fairly close to Husserl's peculiar way of approaching some of the issues we have discussed above. In particular, Fine's elaboration of the distinction between an 'indicative' style and an 'imperatival' style of postulation, together with his emphasis on the latter, should be taken seriously into account when one tries to understand and evaluate Husserl's peculiar attitude which, as we saw, tends systematically to mix and intertwine discourse at the level of theories (axiomatic systems) *and* discourse at the level of manifolds. The reason underlying this attitude is that certain problems he was dealing with could be better expressed at the level of theories (e.g. the problem *under which conditions a theory can be taken to to be saturated*), while others could – at that time – be *more naturally* formulated at the other level (e.g. the question concerning *the expansion of the domain of a certain manifold with new objects*). Now issues of the latter kind are typical cases where an 'imperatival' style of postulation is at work. Just by way of exemplification: Fine's proposal for a '*procedural postulationism*' envisages only one type of simple postulate or atomic instruction,

[216] *PdA* App. 473; *PoA* 476–477 (our emphasis).

[217] Fine 2005.

[218] Fine 2005, 90.

consisting in a procedure for introducing a single new object into the domain (*Introduction*: !*x*C*x*). This postulate may be read as "introduce an object *x* conforming to the condition C(*x*) if there is not already such an object in the domain, otherwise do nothing!"[219] Next, there are four kinds of complex postulates built up from the simple one and conceived of as multiple applications of the simple postulate: "A complex postulate ... requires that we successively, or simultaneously, apply the simple procedures to yield more and more complex extensions of the given domain."[220] Thus in Fine's approach we find the specification of certain 'dynamical' means or procedures for going from given structures to new ones, rather than the specification of certain 'static', axiomatic conditions for characterizing structures as in the 'indicative' style in which a mathematical theory is usually set up.

Let us now consider what Husserl says in the concluding paragraphs of our text:

> We ... must distinguish two things:
>
> 1. Expansion by means of operations and forms of relations. This is ruled out.
> 2. Expansion by means of existential axioms, and thus expansion of the domain (within the sphere of the same operations):
>   α. Defined completely (*perfekt*) while keeping the domain closed in virtue of the existents already defined.
>   β. Defined completely without qualification, even if the expansion of the domain is permitted. Expansion is no longer possible.
>   Again two cases:
>
> a. The axiom-system is to be identically retained.
> b. The axiom-system is preserved only for the old domain. But new objects are defined, and an axiom-system is constructed in such a way that when restricted to the old domain it becomes the old axiom-system. But completeness (*Perfektion*) in the sense that such an expansion must not be possible is not required.[221]

Thus, given a theory T, two kinds of extensions are, under suitable conditions, possible.

(i) We add new axioms written in the old language (these new axioms determine new properties for operations and relations of the old domain). This kind of expansion is possible if, and only if, the corresponding theory is not *saturated*.
(ii) We add existential axioms, i.e. axioms concerning the existence of new individuals, and possibly new operations, such as "there is an *a* such that ...". In other words, we genuinely expand the domain underlying T.

The latter kind of expansion – which is very close to what usually happens in current mathematical practice – has clearly to do with *performing certain operations* on structures. On our reading Husserl is thinking here of applying something similar to Fine's introduction postulate: "Introduce new objects into the domain!"

[219]Loc. cit.
[220]Fine 2005, 91.
[221]*PdA* App. 473–474; *PoA* 477.

3.16 Concluding Remarks

This chapter was devoted to Husserl's *Doppelvortrag* and to Husserl's reappraisal of its key notions in his later *Formal and Transcendental* Logic. In the *Doppelvortrag* he introduces two different concepts of definiteness, absolute definiteness and relative definiteness. We tried to provide an adequate formal candidate for each of them and to challenge several contributions to the topic in the secondary literature. We also tried to give an account of a number of other related – and often rather elusive – issues, whose importance lies in the fact that they have a clear metalogical flavour and that they have a bearing on fundamental questions concerning the nature of the axiomatic method and the foundations of mathematical theories.

We interpreted relative definiteness as syntactic completeness of a theory and absolute definiteness as a closure condition analogous to the axiom of completeness that Hilbert stated for the system of real numbers. 'Absolutely definite' theories are, in our eyes, categorical theories. We then argued against the interpretation of *relative* definiteness (a) as categoricity and (b) as syntactic completeness *restricted to a specified set of formulae*, as well as against the interpretation of *absolute* definiteness as syntactic completeness. We also contested the claim that in the *Doppelvortrag* Husserl confronts the question of the semantic completeness of a theory.

A definite manifold is a set of models structurally very similar to each other. How are we to understand this structural similarity? To solve this problem we reconstructed a path (already explored in the literature) towards an extensional characterization of Husserl's theory of manifolds. It turned out that the correspondence between definite theory and definite manifold that Husserl dreamt of must remain a dream because of certain facts that are acknowledged in current logic.

3.17 General Conclusion

Husserl insistence (in §24 of the third *Logical Investigation*) on defining concepts wherever possible "with mathematical exactness" has opened a fruitful direction of research. Examples thereof are Peter Simons' formalization of Husserl's part-whole theory as well as Casari's and Kit Fine's attempts and Null & R. Simons' formal reconstruction of Husserl's theory of manifolds. Hopefully, this book has also contributed to "the progress" – as Husserl puts it in that *Investigation* – "from vaguely formed, to mathematically exact, concepts and theories."

We tried to address conceptual and historical issues so far almost entirely neglected in the literature and to reconsider well-known issues such as e.g. the Frege–Husserl dispute on the nature of logic, the practice of defining a concept by defining its extension and the true sense of numerical assertions.

We first focused on Husserl's youthful work, the *Philosophy of Arithmetic*. Here we took three topics to be especially worthy of serious consideration: (1) Husserl's definition of number by *Cantorian* abstraction; (2) the controversy between Frege and Husserl concerning the definition of cardinal numbers; (3) issues bearing on the very concept of 'arithmetical operation' and, more generally, on the notion of computational process.

In particular, Husserl's conception of arithmetical operations as procedures of numerical construction as well as *procedures for reducing* complex numerical expressions to the corresponding 'normal forms' turned out to be surprisingly innovative. Even more striking is the fact that Husserl is the first mathematician-philosopher to reflect upon and systematically investigate the problem of circumscribing the totality of computable numerical operations. Our result [Appendix 1] concerning the extensional equivalence of the class of operations Husserl had in mind with the class of partial recursive functions can be seen, or so we argued, as a step in the direction of defining, wherever possible, all concepts "with mathematical exactness".

We saw that Husserl clearly endorsed a version of the thesis that algorithmic signs have a meaning that is exhausted by the operational rules which constitute the algorithm. However, as Husserl stresses (in opposition to George Boole), the logical soundness of the signitive structure has to be guaranteed by the parallelism between an underlying 'conceptual' system and the system of signs. We tried to clarify this epistemological stance by analyzing several texts from the *Nachlass* (dating around 1891–1896) in which Husserl confronts the algebraic approach of Boole and Schröder.

At this point it remains a topic for future research to fill a gap in the history of the notion of computation: how exactly are these early Husserlian reflections connected - *via* Hilbert and Göttingen – with the beginnings of the combinatorial approach to logic as developed by Schönfinkel and Curry around 1924–1928?

Another purpose of the present research has been that of bringing to light Bolzano's often neglected role in Husserl's reflections since 1894–96. Ever so often Husserl works, without explicitly acknowledging it, with Bolzanian distinctions. At the time of the *Prolegomena* Husserl's "logical universe" is a universe of abstract logical contents, such as concepts and propositions. Pure logic studies, here as in Bolzano's monumental *Wissenschaftslehre*, the interconnections between such meaning-entities. In Husserl we noted a certain oscillation (absent from Bolzano) concerning laws vs rules and derivations vs etiological proofs. Husserl is sensitive to the problem of *Methodenreinheit* as we called it: a good mathematical proof should avoid a *metabasis eis allo genos*. But he is also looking for a concept of proof that comports well with common mathematical practice, and that makes for some unclarity.

Already in his *Philosophy of Arithmetic* Husserl talks of an 'Arithmetica Universalis', i.e. a set of calculation rules common to all numerical systems. This notion is further elaborated in Husserl's *Doppelvortrag* of 1901. Here he makes a distinction between the rules of calculation in 'general forms of operations' that belong to universal arithmetic and 'specific forms of operations' that determine a particular field: the latter may be admissible in one field without making sense in another. The 'problem of the imaginary', as Husserl calls it, can be recast as that of finding the conditions that make it possible to extend consistently a given (numerical) deductive theory by new 'specific forms of operations', in order to answer the following question: How is it possible that in calculations the application of 'meaningless forms' yields meaningful and correct results? In order to solve this problem Husserl introduces and studies various interesting properties of, and relations between, axiom-systems and manifolds at the level of structures. In particular, he circumscribes a property of definiteness (relative vs absolute) which he contrasts with notions elaborated at that time by Hilbert (notably, his axiom of completeness).

It was not an easy task to disentangle the intricate and often rather sketchy conceptualizations of the *Doppelvortrag*. Despite some recent contributions a convincing in-depth account and formal reconstruction of the whole range of Husserl's views in this area has yet to be written. It could be a highly significant contribution to the history of the origins and early developments of metamathematics. We just made a beginning with this. A truly comprehensive account would confirm, or so we believe, the late Husserl's often repeated claim to be the father of several important ideas that were subsequently adopted, without acknowledgement, in the logical investigations of Hilbert's school. But this is a story for another occasion.

Bibliography

Works by E. Husserl

Philosophie der Arithmetik. Logische und psychologische Untersuchungen, Bd. I, Pfeffer, Halle a.d.S. 1891. English translation: Willard D (ed) (2003) Philosophy of arithmetic, psychological and logical investigations with supplementary texts from 1887–1901. Kluwer, Dordrecht

Logische Untersuchungen, Bd I: Prolegomena zur reinen Logik, Max Niemeyer, Halle, 1900 (^{2}1913). Bd II: Untersuchungen zur Phänomenologie und Theorie der Erkenntnis, Max Niemeyer, Halle a.d.S, 1901 (^{2}1913–1921). English translation: Findlay JN (ed) (1970) Logical investigations. Routledge, New York

Ideen zu einer reinen Phänomenologie und phänomenologischen Philosophie, Jahrbuch für Philosophie und phänomenologische Forschung I, 1913, 1–323 [as a separate volume: Niemeyer, Halle a.d.S 1913; ^{2}1922; ^{3}1928]

English translation: Ideas pertaining to a pure phenomenology and to a phenomenological philosophy. First Book: Kersten F (1982) General introduction to a pure phenomenology. In: Edmund Husserl, Collected works vol. II. Springer, Berlin

Jahrbuch für Philosophie und phänomenologische Forschung X (1929) Formale und transzendent'xale Logik. Versuch einer Kritik der logischen Vernunft [as a separate volume: Niemeyer, Halle a.d.S 1929]

* * *

Husserliana, Gesammelte Werke. I–XXVI: Martinus Nijhoff, Den Haag 1950–; XXVII–XXXVII: Kluwer, Dordrecht 1989–; XXXVIII–: Springer, New York, 2005

Vol. III: Ideen zu einer reinen Phänomenologie und phänomenologischen Philosophie. Erstes Buch; Allgemeine Einführung in die reine Phänomenologie. Neue, auf Grund der handschriftlichen Zusätze des Verfassers erweiterte Auflage (W. Biemel, ed.), 1950

Vol. XII: Philosophie der Arithmetik. Mit ergänzenden Texten (1890–1901) (L. Eley, ed.), 1970

Vol XVII: Formale und transzendentale Logik. Versuch einer Kritik der logischen Vernunft (P. Jannsen, ed.), 1974

Vol. XVIII: Logische Untersuchungen. Erster Band: Prolegomena zur reinen Logik (E. Holenstein, ed.), 1975

Vol. XIX/1: Untersuchungen zur Phänomenologie und Theorie der Erkenntnis. Erster Teil (U. Panzer, ed.), 1984

Vol. XIX/2: Untersuchungen zur Phänomenologie und Theorie der Erkenntnis. Zweiter Teil (U. Panzer, ed.), 1984

Vol. XXII: Aufsätze und Rezensionen (1890–1910) (B. Rang, ed.), 1979

Husserliana, Materialienbände. I–VII – : Kluwer, Dordrecht 2001 – ; VIII – : Springer, New York 2006
Vol I: Logik: Vorlesung 1896 (E. Schuhmann, ed.), 2001
Vol II: Logik Vorlesung 1902/03 (E. Schuhmann, ed.), 2001

Works by Other Authors

Bolzano B (1810) Beyträge zu einer begründeteren Darstellung der Mathematik. Erste Lieferung. Caspar Widtmann, Prague. Reprint: Wissenschaftliche Buchgesellschaft, Darmstadt 1974

Bolzano B (1969) Wissenschaftslehre. Versuch einer ausfürlichen und größtentheils neuen Darstellung der Logik mit steter Rücksicht auf deren bisherige Bearbeiter, Bd 1–4, Seidel, Sulzbach 1837. In: Berg J et al (ed), Bernard Bolzano-Gesamtausgabe [BGA], Frommann (Holzboog), Stuttgart-Bad Cannstatt 1969 ff, series 1, vols 11–14Translation of selected parts: George R (ed, trans.) (1972) Theory of science. Oxford and Berkelay/Los Angeles

Bolzano B (1975) Paradoxien des Unendlichen (post., Leipzig 1851), Meiner, Hamburg. English translation: Russ S (2004) The mathematical works of Bernard Bolzano. Oxford University Press, Oxford

Bolzano B, Reine Zahlenlehre, in: BGA 2A, 8

Bolzano B, Lehrbuch der Religionswissenschaft, ein Abdruck der Vorlesungshefte eines ehemaliger Religionslehrers an einer katholischen Universitaet, von einigen seiner Schuler gesammelt und herausgegeben, Sulzbach, 1834. Drei Theile in 4 Bände [Kritische neu Ausgabe in: BGA(1 6-8)]

Boole G (1847) The mathematical analysis of logic. Macmillan, Cambridge

Boole G (1854) An investigation of the laws of thought. Macmillan, Cambridge

Brentano F, Psychologie vom empirischen Standpunkt. Duncker & Humblot, Leipzig 1874. Second edition (ed: Kraus O), Meiner, Leipzig 1924

Cantor G, Über unendliche lineare Punktmannigfaltigkeiten, 1–6, Mathematische Annalen 15 (1879), 1–7; 17 (1880), 355–358; 20 (1882), 113–121; 21 (1883), 51–58 and 545–586; 23 (1884), 453–488

Cantor G (1883) Grundlagen einer allgemeinen Mannigfaltigkeitslehre. Teubner, Leipzig

Cantor G, Mitteilungen zur Lehre vom Transfiniten, Zeitschrift für Philosophie und philosophische Kritik, 91 (1887), 81–125; 92 (1888), 240–265

Cantor G (1991) Briefe (eds: Meschkowski H, Nilson W). Springer, Berlin

Cantor G, *Beiträge zur Begründung der transfiniten Mengenlehre* I, II, Mathematische Annalen 46 (1895), 481–512; Mathematische Annalen 49 (1897), 207–246. English translation: Cantor G (1955) Contributions to the founding of the theory of transfinite numbers (ed. & trans: Jourdain Ph E). Dover, New York

Cantor G (1932) *Gesammelte Abhandlungen mathematischen und philosophischen Inhalts* (ed: Zermelo E). Springer, Berlin

Dedekind R (1872) Stetigkeit und irrationale Zahlen. Vieweg, Braunschweig (reprint: Vieweg & Sohn, Braunschweig 1969)

Dedekind R (1888) Was sind und was sollen die Zahlen. Vieweg, Braunschweig 1888 (reprint: Vieweg & Sohn, Braunschweig 1969). English translation: Dedekind R (1963) Essays on the theory of numbers (ed & trans: Beman WW), Dover, New York 1963

Dedekind R, Letter to Keferstein (1890), in: Van Heijenoort, 1997

Dedekind R (1931) Gesammelte mathematische Werke (eds: Fricke R, Noether E, Ore O). Vieweg, Braunschweig

Frege G (1969) Nachgelassene Schriften (eds: Hermes H et al). Meiner, Hamburg

Frege G (2007) Begriffsschrift und andere Aufsätze. Mit E. Husserl und H. Scholz' Anmerkungen (ed: Angelelli I). Olms, Hildesheim

Frege G (1881) Booles rechnende Logik und die Begriffsschrift. In: Frege 1969, 9–52
Frege G (1882) Ueber den Zweck der Begriffsschrift. In: Frege 2007, 97–106
Frege G (1879) Begriffsschrift – Eine der arithmetischen nachgebildete Formelsprache des reinen Denkens. Nebert, Halle a.d.S.
Frege G (1884) Die Grundlagen der Arithmetik. Eine logisch-mathematische Untersuchung über den Begriff der Zahl. Köbner, Breslau
Frege G (1885) Ueber formale Theorien der Arithmetik. In: Frege 1990, 103–111
Frege G (1891) Function und Begriff und Gegenstand. Pohle, Jena
Frege G (1892) Über Begriff und Gegenstand. Zeitschrift für Philosophie und philosophische Kritik 16:192–205
Frege G, Grundgesetze der Arithmetik, begriffschriftlich abgeleitet, Band 1, Pohle, Jena 1893; Band 2, Jena 1903
Frege G (1899) Ueber die Zahlen des Herren H. Schubert. In: Frege 1990, 240–261
Frege G (1894) Rezension von: E. G. Husserl, Philosophie der Arithmetik, I. In: Frege 1990, 179–192
Frege G (1918) Der Gedanke, in: Beiträge zur Philosophie des Deutschen Idealismus I/2
Frege G (1923) Gedankengefüge. In: Beiträge zur Philosophie des Deutschen Idealismus III/1
Frege G (1976) Wissenschaftlicher Briefwechsel (eds: Gabriel G et al). Meiner, Hamburg
Grassmann H (1844) Die Wissenschaft der extensiven Größe oder die Ausdehnungslehre, eine neue mathematische Disciplin dargestellt und durch Anwendung erläutert. I. Teil. Die lineale Ausdehnungslehre ein neuer Zweig der Mathematik dargestellt und durch Anwendungen auf die übrigen Zweige der Mathematik, wie auch auf die Statik, Mechanik, die Lehre von Magnetismus und die Krystallonomie erläutert. Wiegand, Leipzig
Hankel H (1867) Theorie der complexen Zahlensysteme. Voss, Leipzig
Hilbert D (1899) Grundlagen der Geometrie. Teubner, Leipzig
Hilbert D (1900a) Über den Zahlbegriff. Jahresbericht der Deutschen Mathematiker-Vereinigung 8:180–184
Hilbert D (1900b) Mathematische Probleme. Nachrichten von der Königlichen Gesellschaft der Wissenschaften zu Göttingen 3:253–297. Reprint in: Hilbert D, 1965
Hilbert D (1905) Über die Grundlagen der Logik und der Arithmetik, in: Verhandlungen des Dritten Internationalen Mathematiker-Kongresses in Heidelberg vom 8. bis 17. August 1904, Teubner, Leipzig, pp 174–185. English translation: van Heijenoort J, 1967, 129–138
Hilbert D (1918) Axiomatisches Denken. Mathematische Annalen 78:405–415. Reprint in: Hilbert D, 1965
Hilbert D (1923) Die Logische Grundlagen der Mathematik. Mathematische Annalen 88:151–165
Hilbert D (1926) Über das Unendliche. Mathematische Annalen 95:161–190
Hilbert D (1935) Gesammelte Abhandlungen. Dritter Band: Analysis, Grundlagen del Mathematik, Physik, Verschiedenes. Springer, Berlin (Reprint: AMS Chelsea Publishing, New York 1965)
Kerry B, Über Anschauung und ihre psychische Verarbeitung (I–VIII), Vierteljahrsschrift für wissenschaftliche Philosophie, IX (1885) – XVI (1891)
Kraus O (1919) Franz Brentano. Zur Kenntnis seines Lebens und seiner Lehre, Beck, München (includes: E. Husserl, Erinnerungen an Franz Brentano)
Leibniz GW (1684) Meditationes de cognitione, veritate et ideis. In: Akademieausgabe, VI.4A, N. 141, Akademie-Verlag, Berlin 1999
Leibniz GW (1687) Non inelegans specimen demonstrandi in abstractis. In: Akademieausgabe, VI.4A, N. 178, Akademie-Verlag, Berlin 1999
Leibniz GW (1704) Nouveaux Essais sur l'entendement humain. In: Akademieausgabe, VI.6, Akademie-Verlag, Berlin 1962
Noether E, Cavaillès J (eds) (1937), G. Cantor – R. Dedekind Briefwechsel. Hermann, Paris
Schröder E (1873) Lehrbuch der Arithmetik und Algebra. Teubner, Leipzig
Schröder E (1877) Der Operationskreis des Logikkalkuls. Teubner, Leipzig

Schröder E (1898) Über zwei Definitionen der Endlichkeit und G. Cantorsche Sätze. Nova Acta Academiae Caesareae Leopoldino-Carolinae Germanicae Naturae Curiosum 71
Stolz O (1881) B. Bolzano's Bedeutung in der Geschichte der Infinitesimalrechnung. Mathematische Annalen 18:518–519
Stolz O, Vorlesungen über die allgemeine Arithmetik, Teubner, Leipzig 1885 (vol 1), 1886 (vol 2)
Tarski A (1925) Sur les ensembles finis. Fundamenta Mathematicae 6:45–95
Tarski A (1951) A decision method for elementary algebra and geometry (2nd edn rev.). University of California Press, Berkley, CA
Weierstrass K (1966) Vorlesungen zur Funktionentheorie (summer semester 1874), elaborated by G. Hettner. In: Behnke H, Kopfermann KL (eds) Festschrift zur Gedächtnisfeier für Karl Weierstrass 1815–1965, Wissenschaftliche Abhandlungen der Arbeitsgemeinschaft für Forschung des Landes Nordrhein-Westfalen. Westdeutscher Verlag, Köln
Weyl H (1910) Über die Definitionen der mathematischen Grundbegriffe. In: Weyl H (1968) Gesammelte Abhandlungen (ed: Chandrasekharan K), Band 1. Springer, Berlin, pp 298–304

Studies

Abrusci VM (ed) (1978) David Hilbert. Ricerche sui fondamenti della matematica. Bibliopolis, Napoli
Abrusci VM (1981) "Proof", "theory", and "foundations" in Hilbert's mathematical work from 1885 to 1900. In: Dalla Chiara ML (ed) Italian studies in the philosophy of science. Kluwer, Dordrecht, pp 453–491
Awodey S, Reck EH (2002) Completeness and categoricity. Part I: Nineteenth-century axiomatics to twentieth-century metalogic. Hist Philos Logic 23:1–30
Beaney M (ed) (1997) The Frege Reader. Blackwell, Oxford
Behboud A (1997) Remarks on Bolzano's collections. In: Künne W, Siebel M, Textor M (eds) Bolzano and analytic philosophy. Rodopi, Amsterdam, pp 109–115
Behboud A (2000) Bolzano's Beiträge zur Mathematik und ihrer Philosophie, Bern Studies in the History and Philosophy of Science, Bern
Bell D (1990) Husserl. Routledge, London/New York
Benoist J (2003) Review of E. Husserl, Logik: Vorlesung 1896. Husserl Stud 19:237–242
Berg J (1962) Bolzano's logic. Almquist & Wiksell, Stockholm
Blanchette PA (1996) Frege and Hilbert on consistency. J Phil 93:317–336
Bucci P (2000) Husserl e Bolzano. Alle origini della fenomenologia. Unicopli, Milano
Buhl G (1961) Ableitbarkeit und Abfolge. Kant-Studien, Ergänzungshefte 83
Buonazia S (1996) Per una formalizzazione della teoria husserliana dell'intero e della parte. Rivista di Filosofia 87:287–313
Cantini A (1979) I fondamenti della matematica da Dedekind a Tarski. Loescher, Torino
Carnap R (2000) Untersuchungen zur allgemeinen Axiomatik. In: Bonk T, Mosterin J (eds) Wissenschaftliche Buchgesellschaft, Darmstadt
Casari E (1973) La filosofia della matematica del '900. Sansoni, Firenze
Casari E (1985) L'universo logico bolzaniano. Rivista di Filosofia 76:339–366
Casari E (1987) Matematica e verità. Rivista di Filosofia 78:329–350
Casari E (1991) Anche la Filosofia dell'Aritmetica compie cent'anni, in: Atti del Convegno: Peano e i fondamenti della matematica. Mucchi, Modena, pp 35–49
Casari E (1992) An interpretation of some ontological and semantical notions in Bolzano's logic. In: Florence Centre (ed), pp 55–105
Casari E (1997a) La critica dello psicologismo. In: Rossi P, Viano CA (eds), Storia della Filosofia, vol 5. Laterza, Roma-Bari, pp 553–552

Casari E (1997b), Sull'origine dell'«oggettivo» in Bolzano. In: Fabris A, Fioravanti G, Moriconi E (eds), Logica e teologia. Studi in onore di Vittorio Sainati. ETS, Pisa, pp 93–115

Casari E (1999) Husserl. In: Rossi P, Viano CA (eds) Storia della Filosofia, Vol. 6/1. Laterza, Roma-Bari, pp 161–173

Casari E (2000) On Husserl's theory of wholes and parts. Hist Phil Logic 21:1–43

Casari E (2004) Logic and the laws of possible being. In: Marsonet M, Benzi M (eds) Logic and metaphysics. Name edizioni, Genova, pp 16–43

Casari E (2007) On the relationship between parts and wholes in Husserl's phenomenology. In: Boi L, Kerszberg P, Patras F (eds) Rediscovering phenomenology: phenomenological essays on mathematical beings physical reality, perception and consciousness. Springer, Berlin, pp 67–102

Cavaillès J, Méthode axiomatique et formalisme (1938). In: Cavaillès 1994, pp 1–202

Cavaillès J, Mathématique et formalisme (1949). In: Cavaillès 1994, 659–664

Cavaillès J (1994) Oeuvres complètes de Philosophie des Sciences. Hermann, Paris

Centrone S (2005) Husserl on the concepts of 'operation' and 'system of operations'. Proceedings of the Husserl Circle at University College Dublin, 35th Annual Meeting (Dublin, June 9–12, 2005). Humanities Institute of Ireland, Dublin, pp 283–299

Centrone S (2006) Husserl on the 'Totality of all conceivable arithmetical operations'. Hist Phil Logic 27:211–228

Church A (1936) An unsolvable problem of elementary number theory, American Journal of Mathematics 58:345–363

Corry L (2004) David Hilbert and the axiomatization of physics (1898–1918). Kluwer, Dordrecht

Da Silva JJ (2000) Husserl's two notions of completeness. Husserl and Hilbert on completeness and imaginary elements in mathematics. Synthese 125:417–438

Dreyfus H (ed) (1982) Husserl, intentionality and cognitive science. MIT Press, Cambridge, MA

Dummett M (ed) (1991a) Frege and the paradox of analysis. In: Frege and other philosophers. Clarendon Press, Oxford, pp 17–52

Dummett M (1991b) Frege – philosophy of mathematics. Duckworth, London

Fine K (1994) Essence and modality. In: Tomberlin J (ed) Philosophical perspectives 8: logic and language. Ridgeview Publishing, Atascadero, pp 1–16

Fine K (1995) Part-whole. In: Smith B, Smith DW (eds) The Cambridge companion to Husserl. Cambridge University Press, Cambridge, pp 463–485

Fine K (1998) Cantorian abstraction: a reconstruction and defence. J Phil 95:599–634

Fine K (2005) Our knowledge of mathematical objects. In: Gendler TS, Hawthorne J (eds) Oxford studies in epistemology: Vol. 1. Clarendon Press, Oxford, pp 89–110

Florence Centre for the History and Philosophy of Science (ed) (1992) Bolzano's Wissenschaftslehre 1837–1987. Olschki, Firenze

Føllesdal D (1958) Husserl und Frege: ein Beitrag zur Beleuchtung der Entstehung der phänomenologischen Philosophie, Aschehoug & Co. English translation (Husserl and Frege: A Contribution to Elucidating the Origins of Phenomenological Philosophy) in Haaparanta L (ed.) Mind, Meaning and Mathematics: Essays on the Philosophical Views of Husserl and Frege. Synthese Library 237, Kluwer, Dordrecht 1994

Føllesdal D, Husserl's conversion from psychologism and the Vorstellung-meaning-reference distinction: two separate issues. In: Dreyfus H 1982, pp 52–56

Føllesdal D, Husserl's conversion from psychologism and the Vorstellung-meaning-reference distinction: two separate issues. In: Dreyfus H 1982, pp 52–56

Føllesdal D (1995) Gödel and Husserl. In: Hintikka J (ed) Essays on the development of the foundations of mathematics. Kluwer, Dordrecht, pp 427–446

Fraenkel A (1919) Einleitung in die Mengenlehre. Springer, Berlin (1923^{2}, 1928^{3})

Frewer M (1981) Felix Bernstein. Jahresbericht der Deutschen Mathematiker-Vereinigung 83:84–95

George R (1972) Enthymematic consequence. Am Phil Quart 9:113–116

George R (1983) Bolzano's consequence, relevance and enthymemes and a postscript on fallacies. J Phil Logic 12:299–318, 319–325

Giusti M (2000) Punti di vista esterni alla Pedagogia. La rilettura di E. Husserl fatta da Ettore Casari. Encyclopaideia 4(7):103–116

Hale B, Wright C (2001) The reason's proper study – essays towards a Neo-Fregean philosophy of mathematics. Clarendon Press, Oxford

Hartimo MH (2007) Towards completeness: Husserl on theories of manifolds 1890–1901. Synthese 156:281–310

Heck R (1993) The development of arithmetic in Frege's Grundgesetze der Arithmetik. J Symbolic Logic 58:579–601

Hintikka J (ed) (1995) From Dedekind to Gödel. Essays in the development of the foundations of mathematics. Kluwer, Dordrecht

Ierna C (2005) The beginnings of Husserl's philosophy, Part I: From Über den Begriff der Zahl to Philosophie der Arithmetik. In: The new yearbook for phenomenology and phenomenological philosophy, V. Noesis Press, Seattle, pp 1–56

Ierna C (2006) The beginnings of Husserl's philosophy, Part 2: Philosophical and mathematical background. In: The new yearbook for phenomenology and phenomenological philosophy, VI. Noesis Press, Seattle, pp 33–81

Kline M (1972) Mathematical thought from ancient to modern times. Oxford University Press, New York

Krickel F (1995) Teil und Inbegriff, Bernard Bolzanos Mereologie. Academia Verlag, St. Augustin

Künne W (2003) Conceptions of truth. Clarendon Press, Oxford

Künne W (2008) Versuche über Bolzano/Essays on Bolzano. Academia Verlag, St. Augustin

Künne W (2009) Die Philosophische Logik Gottlob Freges. Klostermann, Frankfurt/M

Lewis D (1991) Parts of classes. Basil Blackwell, Oxford

Lohmar D (1989) Phänomenologie der Mathematik. Elemente einer phänomenologischen Aufklärung der mathematischen Erkenntnis nach Husserl. Kluwer, Dordrecht

Lohmar D (2000) Edmund Husserls "Formale und transzendentale Logik". Wissenschaftliche Buchgesellschaft, Darmstadt

Majer U (1997) Husserl and Hilbert on completeness: a neglected chapter in early twentieth century foundations of mathematics. Synthese 110:37–56

Mancosu P (1999) Bolzano and Cournot on mathematical explanation. Revue d'histoire des sciences 42:429–455

Miller PJ (1982) Numbers in presence and absence. A study of Husserl's philosophy of mathematics. Martinus Nijhoff, The Hague

Morscher E (2008) Bernard Bolzano's life and work. Academia Verlag, St. Augustin

Mulligan K (2004) Essence and modality: The quintessence of Husserl's theory. In: Siebel M, Textor M (eds) Semantik und Ontologie. Ontos Verlag, Frankfurt, pp 387–419

Mulligan K, On the essence of functions: Bolzano, Kerry and Husserl. Unpublished ms.

Null GT, Simons RA (1982) Manifolds, concepts and moment abstrakta. In: Smith B (ed) Parts and moments, studies in logic and formal ontology. Philosophia Verlag, München, pp 439–479

Ortiz Hill C, Husserl and Frege on Substitutivity (1994a), In: Ortiz Hill C, Rosado Haddock GE 2000b, 1–22

Ortiz Hill C, Frege's attack on Husserl and Cantor (1994b). In: Ortiz Hill C, Rosado Haddock GE 2000b, 95–107

Ortiz Hill C, Husserl and Hilbert on completeness. In: Hintikka J 1995, 143–163

Ortiz Hill C, The varied sorrows of logical abstraction (1997a). In: Ortiz Hill C, Rosado Haddock GE 2000b, 67–96

Ortiz Hill C, Did Georg Cantor influence Edmund Husserl (1997b). In: Ortiz Hill C, Rosado Haddock GE 2000b, 137–160

Ortiz Hill C, *Husserl's Mannigfaltigkeitslehre* (2000a), in: Ortiz Hill C, Rosado Haddock GE 2000b, 161–168

Ortiz Hill C, Abstraction and idealization in Edmund Husserl and Georg Cantor prior to 1895 (2004). In: Ortiz Hill C, Rosado Haddock GE 2000b, pp 109–135

Ortiz Hill C, Rosado Haddock GE (2000b) Husserl or Frege? Meaning, objectivity, and mathematics. Open Court, Chicago
Ortiz Hill C (2002) Tackling three of Frege's problems: Edmund Husserl on sets and manifolds. Axiomathes 13:79–104
Paoli F (1991) Bolzano e le dimostrazioni matematiche. Rivista di Filosofia 83:221–242
Peckhaus V (1990) Hilbertprogramm und Kritische Philosophie. Vandenhoeck & Ruprecht, Göttingen
Picardi E (1994) Kerry und Frege über Bgriff und Gegenstand. Hist Phil logic 15:9–32
Picker B, Die Bedeutung der Mathematik für die Philosophie Edmund Husserls. In: Philosophia Naturalis 7:266–355
Presburger M (1929) Über die Vollständigkeit eines gewissen Systems der Arithmetik ganzer Zahlen, in welchem die Addition als einzige Operation hervortritt, Comptes Rendus du I congrès de Mathématiciens des Pays Slaves, Warszawa 92–101
Purkert W, Ilgauds HJ (1987) Georg Cantor: 1845–1918. Birkhäuser, Basel
Reid C (1970) Hilbert. Springer, Berlin/Heidelberg/New York
Remnant P, Bennett JP (1996) "Introduction", in: *Leibniz GW, New essays on human understanding;* translated and edited by Peter Remnant and Jonathan Bennett. Cambridge University Press, New York, pp vii–xxx
Rigamonti G (ed) (1992) Georg Cantor. La formazione della teoria degli insiemi, Sansoni, Firenze
Robinson J (1950) General recursive functions. Proc Am Math Soc 1:703–718
Rosado Haddock GE (1987) Husserl's epistemology of mathematics and the foundation of platonism in mathematics. Husserl Studies 4:81–102
Scholz H (1930) Die Axiomatik der Alten. In: Mathesis Universalis, Wissenschaftliche Buchgesellschaft, Darmstadt 1969, pp 27–44
Schuhmann E, Schuhmann K (eds), Husserls Manuskripte zu seinem Göttinger Doppelvortrag von 1901. Husserl Studies 17:87–123
Schuhmann K (ed) (1977) Husserl-Chronik. Denk- und Lebensweg Edmund Husserls, Husserliana-Dokumente, vol I. Martinus Nijhoff, Den Haag
Sebestik J (1992) Logique et mathématique chez Bernard Bolzano. Vrin, Paris
Siebel M (1996) Der Begriff der Ableitbarkeit bei Bolzano. Academia Verlag, St. Augustin
Sieg W (2002) Beyond Hilbert's reach. In: Malament D (ed) Reading natural philosophy – essays in the history and philosophy of science and mathematics. Open Court Press, Chicago, pp 363–405
Simons PM (1982) Three essays in formal ontology. In: Smith B (ed), 111–260
Simons PM (1997) Bolzano on collections. In: Künne W, Siebel M, Textor M (eds) Bolzano and analytic philosophy. Rodopi, Amsterdam, pp 87–108
Simons PM (2007) What numbers really are. In: Auxier RE, Hahn LE (eds) The philosophy of Michael Dummett. Open Court, La Salle, pp 229–247
Simons PM, What sets are. Unpublished ms.
Smith B (ed) (1982) Parts and moments. Philosophia, Münschen
Smith B, Smith DW (eds) (1995) The Cambridge companion to Husserl. Cambridge University Press, Cambridge
Stein H (1988) Logos, logic, and logistiké: some philosophical remarks on the 19th-century transformation of mathematics. In: Aspray W, Kitcher PH (eds), History and philosophy of modern mathematics. Minnesota Studies in the Philosophy of Science, vol XI. University of Minnesota Press. Minneapolis, MN, pp 238–259
Sundholm G (ed) (1997) Inference versus consequence. In: Logica yearbook 1997. Filosofia Publishers, Czech Academy of Science, Prague, pp 26–35
Tatzel A (2002) Bolzano's theory of ground and consequence. Notre Dame J Formal Logic 43:1–25
Tieszen R (1990) Frege and Husserl on Number. Ratio (New Series) 3:150–164
Tieszen R (1994) The philosophy of arithmetic: Frege and Husserl. In: Tieszen R 2005, 314–336
Tieszen R (1995) Mathematics and transcendental phenomenology. In: Tieszen R 2005, 46–68

Tieszen R (1996) Logicism, Impredicativity, Formalism. Some remarks on Poincaré and Husserl. In: Tieszen R 2005, 294–313

Tieszen R (2000) The philosophical background of Weyl's mathematical constructivism. In: Tieszen R 2005, 248–275

Tieszen R (2001) Intuitionism, meaning theory, and cognition, in: Tieszen R 2005, 227–247

Tieszen R (2004) Science as a triumph of the human spirit and science in crisis. Husserls and the fortunes of reason. In: Tieszen R 2005, 21–45

Tieszen R (2005) Phenomenology, logic and the philosophy of mathematics. Cambridge University Press, Cambridge, NY

van Atten M (2002) Why Husserl should have been a strong revisionist in mathematics. Husserl Stud 18:1–18

van Heijenoort J (1967) From Frege to Gödel: a source book in mathematical logic, 1879–1931. Harvard University Press, Cambridge

Webb J (1980) Mechanism, mentalism and mathematics. Reidel, Dordrecht

Willard D (1974) Concerning Husserl's view of number. Southwest J Phil V 3:97–109

Willard D (1984) Logic and the objectivity of knowledge: a study in Husserl's philosophy. Ohio University Press, Athens

Yasuhara A (1971) Recursive function theory and logic. Academic Press, New York

Author Index

Subject Index

FSC
www.fsc.org
MIX
Papier | Fördert gute Waldnutzung
FSC® C083411

Kolibri 360 GmbH
Ferdinand-Jühlke-Straße 7
99095 Erfurt, Deutschland
produktsicherheit@kolibri360.de